# Regulatory and Advanced Regulatory Control:
## Application Techniques

by David W. Spitzer, P.E.

 **Instrument Society of America**

INSTRUMENT SOCIETY OF AMERICA
67 Alexander Drive
P.O. Box 12277
Research Triangle Park
North Carolina 27709


**Library of Congress Cataloging-in-Publication Data**
Spitzer, David W.
  Regulatory and advanced regulatory control : application
techniques / by David W. Spitzer
      p. cm.
  Includes index.
ISBN 1-55617-487-X
1. Automatic control.    I. Title.
TJ213S655  1993
629.8--dc20

# *Dedication*

This work is dedicated to my children, Deborah Goldie and Michael James, in the hope that it will inspire them and others to continue their education and their pursuit of knowledge; and to my wife Ida for understanding why I was so late to bed for so many months.

# *About the Author*

DAVID WILLIAM SPITZER, P.E. is the manager of Utility and Instrumentation Engineering at Nepera, Inc., Harriman, NY, where he is responsible for technical support and direction in the electrical, instrumentation, and utility areas on a plant wide basis. He has proposed many control and energy-saving projects that are in planning, design, construction, and operational stages.

Mr. Spitzer serves on the American Society of Mechanical Engineers' Committee on Measurement of Fluid Flow in Closed Conduits. He is also a member of the Instrument Society of America and the Association of Energy Engineers.

A frequent instructor at the ISA Training Center in Raleigh, NC, and throughout the United States and internationally in the Short Course Program, Mr. Spitzer is the author of *Industrial Flow Measurement*, part of ISA's Resources for Measurement and Control (RMC) Series, Volume Editor of *Flow Measurement*, part of ISA's Practical Guides for Instrumentation and Control series, and *Variable Speed Drives: Principles and Applications for Energy Cost Savings*.

# *Table of Contents*

# The Challenge of Advanced Regulatory Control

Whether the instrumentation control hardware consists of local mechanical, pneumatic, or electronic controls, electronic or pneumatic analog controllers, microprocessor-based controllers, programmable logic controllers, personal computer-based controls, or a distributed control system, a keen awareness of advanced regulatory control is required by engineers, technicians, educators, managers, salespersons, and marketing personnel in order that appropriate control technology be applied to industrial processes.

Make no mistake—the forces of "instant gratification" are definitely at work in the wonderful world of process control. It is not uncommon to find support for improved control systems driven by the aesthetics of pretty pictures, "buzz" words, the need to be the "first one on the block" to have the latest thing, or the need *not* to be the only one without one. In these situations, little attention is truly focused on the quality and sophistication of control.

The quality and sophistication with which control strategies are developed and implemented varies with the skill of the process control engineer. This skill is heightened by a strong control engineering education, exposure to skilled process control engineers, specialization in the field of process control, and intimate knowledge of the process.

Many persons perform process control engineering on a part-time basis without the benefit of a process control education, exposure to skilled practitioners, or understanding of the process. As verified by the observation of numerous control strategy implementations, the typical result of this approach is a control system that "works."

Such a situation can be illustrated by the author's experience with an individual who implemented a distributed control system and enthusiastically described it, especially with regard to its flexibility and ease of configuration. After responding with silence to questions that probed the sophistication of the control strategies, this individual sat in awe as the operation of a cascade control loop was described. Despite its outstanding functional capability, this individual's distributed control system implements less-sophisticated control than is possible with a pneumatic analog control system. Nonetheless, it cannot be denied that the distributed control system "works," that the plant can boast of having installed the latest and greatest technology, and that this experienced "expert" may influence the decisions of others who are less knowledgeable than he.

This type of implementation occurs because of a series of events that typically begins with management's recognition and support for improved process control. Due to the skill level of the person selected to perform the control system engineering function, inappropriate or rudimentary control strategies may be implemented, which result in the loss of an opportunity to improve plant efficiency. Operators have little choice but to accept the control system and work around its shortcomings, including operating in the manual mode (system abandonment) because the controls are inconvenient to operate or do not function properly. However, despite these shortcomings, the operators are usually satisfied because the installation represents improvement.

The control system that "works" may be regarded as a success, primarily because management is rarely versed in process control technology, has difficulty judging the technical quality of the installation, and is usually forced to draw conclusions about instrumentation from the opinions of others who are not skilled control system engineers. It is not uncommon for an experienced control system engineer to review such an installation and recognize the need for an additional control system upgrade to improve process performance and personnel safety.

Were the problem solely caused by errors, improved process control and safety could be achieved with a few more expenditures to correct deficiencies. However, the underlying problem is that an installation that "works" is deemed acceptable, regardless of whether it truly is. What is not recognized is that the process may really be poorly controlled and/or unsafe because of inappropriate control strategies and interlocks.

The challenge of providing appropriate controls for industrial processes is a formidable task due to the many facets, details, and idiosyncrasies of industrial processes and the instruments selected to be integrated into their operation. Virtually anyone, with or without a process control background, can put together a control system that "works" with the help of a telephone and enough vendors; however, to take advantage of technology to improve the process through the innovative application of instrumentation requires a skilled, process control system engineer.

A skilled process control system engineer can promote process improvements based upon specific knowledge of the instrumentation and the process. Such improvements can streamline the process, provide safe operation, improve efficiency, eliminate the need for equipment, and reduce installation, operation, and maintenance costs. Despite claims to the contrary, individuals who have not specialized in process control and have not exposed themselves to more sophisticated control techniques have not generally produced innovative installations that exploit the potential of control technology and open the door to additional process improvement.

Most industrial control is performed by utilizing a combination of manual and regulatory control. Despite the installation of additional field transmitters and advanced control capabilities, regulatory controls that emulate conventional pneumatic and electronic analog controllers abound. To become more productive and to improve safety, advanced controls must be appropriately and correctly applied to industrial processes.

Traditionally, for lack of a better definition, advanced control has been defined as continuous control that was more complex than regulatory control (containing one measurement device, one controller, and one final control element). Advanced regulatory control can be segregated from advanced control strategies such as model-based control, state variable control, optimal control, dynamic matrix control, statistical process control, and the like.

Herein, advanced regulatory control encompasses continuous controls that are more complex than regulatory controls (as defined above), but still retain the measurement, the controller, and the final control element format of regulatory control. The complexities of advanced regulatory control may be due to multiple measurements, controllers, and/or final control elements as well as logical manipulation of the control loop.

Advanced regulatory control is but one tool in the process control system engineer's arsenal; however, advanced regulatory control can and should be applied to the majority of control systems. In some circles, it is believed that if advanced regulatory control were properly and appropriately applied, the need to resort to applying advanced controls would be greatly diminished.

It should be noted that some aspects of advanced regulatory control have been practiced for decades using mechanical, pneumatic, and electronic instrumentation. Distributed control systems and microprocessor-based controllers have dramatically reduced the cost of performing advanced regulatory control functions and have simplified their implementation.

This book is intended to describe advanced regulatory control and its application to continuous processes in a nonmathematical format and in as practical a manner as possible in order to be of benefit to all skill levels. Manual and regulatory control are described in the text as a prelude to advanced regulatory control so that their differences might be explored.

Discussions of control system hardware are avoided, because the *control function*, not its hardware implementation, is what is important in improving control of the process. This contrasts the importance of the measurement instruments and final control elements, which, if not properly selected and installed, can cause the control system to operate in the GIGO mode, that is, "garbage in, garbage out." In this mode, and despite control system sophistication, poor process measurements and/or poor process manipulation methods hamper control of the process. In many cases, poor process measurements may not show process instability, which would lead the uninitiated to the erroneous conclusion that the control is adequate and the instrumentation is not in need of upgrading.

This text focuses on control but integrates the operation of measurement devices and final control elements into discussions, because their importance is often the limiting factor in control system performance.

# ❖ Chapter 1

# *Introduction*

People say that the world is getting smaller, in large part due to technology, yet ironically technology is becoming so specialized that there are often two or more persons with distinctly different job functions performing the same function as one person did only a few years ago. It would seem logical that a person regularly practicing a particular skill will perform better and more efficiently than someone who does not. However, as these job functions migrate apart, there must be interfaces between each person's work product that can be difficult to maintain.

The field of instrumentation has not escaped this trend. Ignoring the specialists (such as in the areas of analyzers, programmable logic controllers, flow measurement, control valves, and the like) that may be employed by larger organizations, the primary disciplines responsible for the design and construction of instrumentation projects are instrumentation engineers who specify and supervise the installation of field hardware, and process control engineers who develop control strategies and implement them in software. This separatist trend has also permeated the literature where one finds treatises on narrow subjects that are separated by discipline.

For efficient and safe plant operation, the instrumentation and process control *systems* must work together. As instrumentation and process control continue to drift farther apart, problems are created wherein the disciplines work in a vacuum—a situation that can result in a loss of opportunity. Under these conditions, it is not uncommon to find instrument engineers who do not understand the concept of object-oriented programming, nor must one wander very far to find process control engineers who do not understand how instruments work and what they measure.

This book attempts to bridge this gap by addressing the control problem in the context of the control *system*, including the integration of field instrumentation. For this reason, this book should be of benefit to engineers, technicians, and educators of all skill levels.

This book does not teach specific selection and implementation skills, but rather provides the reader with the underlying technical understanding to implement a control *system* and highlights some of the pitfalls. The evolutionary journey flows from manual control to regulatory control to advanced regulatory control, with each successive subject building on previously covered material. Therefore, it is strongly recommended that the text be read in its entirety to fully grasp the concepts that are presented.

# ❖ Chapter 2
# *Manual Control*

Manual control, in which a minimum of local measurements and manual throttling valves are required, is becoming increasingly rare in its pure form even though it offers some advantages. The devices required for manual control are inexpensive relative to automated instrumentation and are easy to specify and install. Piping requirements are reduced by eliminating bypass valve arrangements and pneumatically operated equipment, while electrical requirements are reduced by installing local indication only.

Incorrect installation techniques and inappropriate instrument selection for manual control can be masked by the lack of process measurements and the operator's ability to compensate for these hindrances. Manual control is perhaps the most flexible of all control systems in that changes can be made immediately by simply telling the operator to make them.

Control systems engineering is typically not done for manual systems because there is virtually no control system to design, and what can be designed is so rudimentary and forgiving that virtually anyone can, and does, perform the function.

Upon closer examination, however, many apparent advantages of manual control can quickly be seen as disadvantages. Manual control is less expensive to install than other control systems, but more operators are required to operate the system because a number of adjustments must be made at the same time in different locations, especially when the safety of the process is a concern. This approach exposes the operator to potential hazards because a person must be physically present at the equipment for adjustments to be made. In addition, the actions of one operator may unknowingly affect another operator's part of the process, and create a process upset that could cause a potentially unsafe condition (aside from economic penalties).

## Example 2-1:

Consider the case of an operator who pumps a liquid from one area of the plant to another. A second operator in the other area adjusts a valve that modulates the liquid flow into the process. Seeing that the supply tank level is falling, the first operator may adjust a valve in the first area that reduces the flow to allow the level to rise. Meanwhile, the second operator may be in an upset condition requiring maximum liquid flow. The manual control strategies of these operators is (at least temporarily) in conflict and could cause a safety hazard if control of the second process is lost. Further, this type of incident can occur within the same area of the plant.

System flexibility may appear to be an advantage; however, it quickly becomes a disadvantage when one considers that, in a 24-hour-a-day operation, all four operators and their substitutes must be told of any and every change. Further, it is unlikely that all four operators perform their normal functions in exactly the same way, which could result in variations in process conditions and product quality.

From a technical perspective, upgrading manual operations can provide significant opportunities for process improvements, capacity increases, and increased productivity. However, if the system works, specific justification for the expenditures is usually difficult to find prior to installation, especially since the skilled process control engineer usually has no choice but to scrap virtually all of the existing instruments because of their age and lack of versatility.

Control system expenditures are often justified on the basis of improved quality and increased productivity that are expected when the process is maintained at a given operating condition. Yield improvements and capacity increases can occur by observing and adjusting operating conditions, but the amount of improvement often cannot be predicted prior to installation.

## Example 2-2:

The manually controlled boiler shown in Figure 2-1 is operated to control the steam header pressure to the plant. Without the clutter of the required flame management system on the figure, it can be seen that a minimum of instrumentation has been installed: two pressure gages, a sight glass, two throttling valves, and a damper.

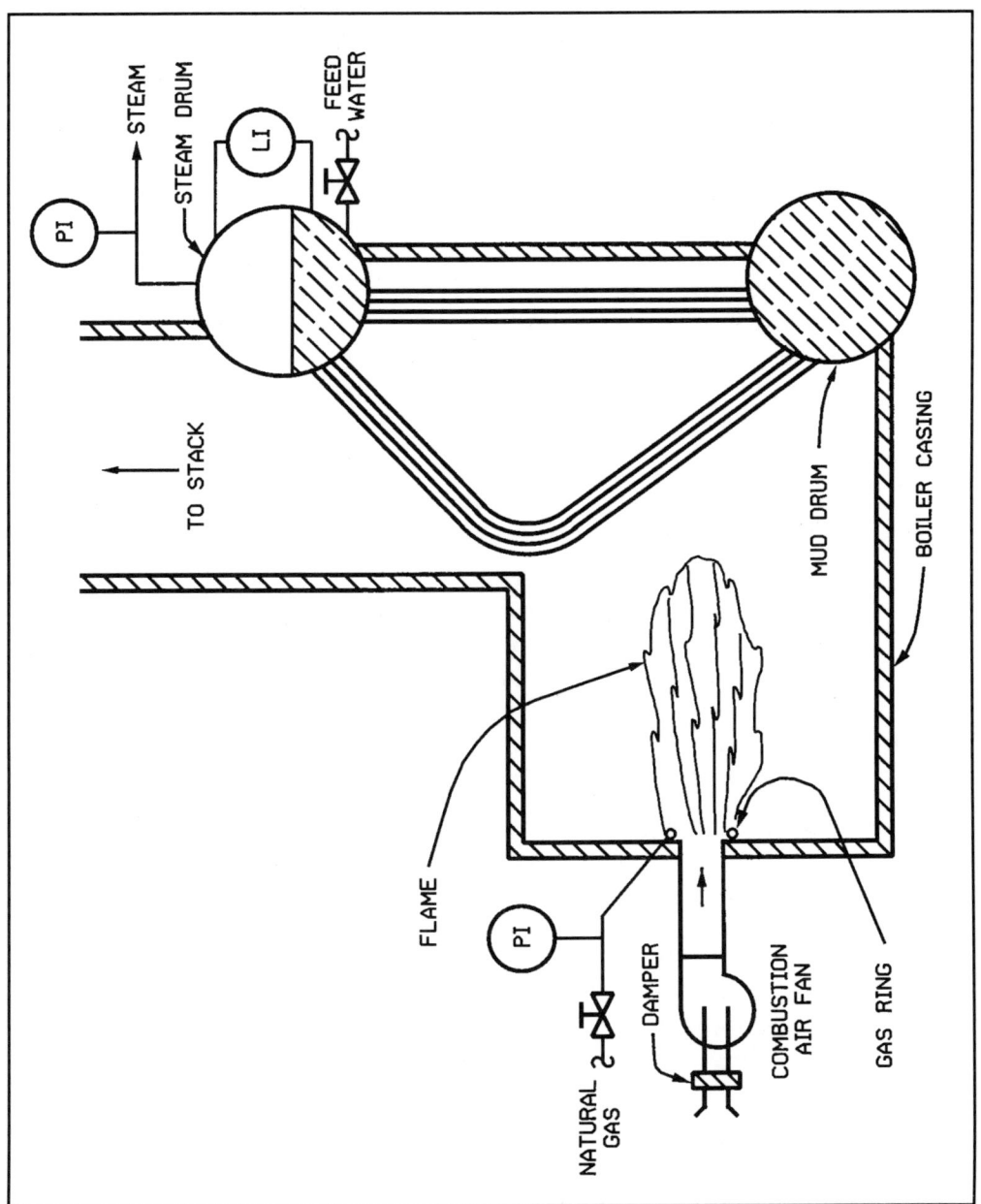

**FIG. 2-1.** Manually controlled boiler (flame management system not shown)

Steam header pressure (indicated at the top of the boiler) can be controlled by simultaneously manipulating the fuel valve and the air damper on the burner. In theory, an operator can perform this control function, but a logistics problem occurs: how can the header pressure be observed while the burner is being adjusted? One solution is to remotely mount the steam header pressure gage near the burner.

The operator makes adjustments to the air and fuel flows by simultaneously adjusting the natural gas valve and the combustion fan damper while visually observing the flame, the natural gas pressure, and the steam drum pressure. The operator must then wait for the effect of the firing change and any concurrent plant load changes on header pressure before making any additional adjustments. Inappropriate adjustment of the air and/or gas can cause a boiler trip due to a flameout (due to an improper mixture) or high steam pressure (due to overfiring for a given steam load). In theory it should be possible for an operator to perform these functions, but in practice the quality with which an operator can perform this function *consistently* is in question.

While the above is occurring, the boiler drum level is maintained by another operator who visually observes the level gage and adjusts the feedwater valve. The drum level will vary in response to steam load changes, burner adjustments made by the first operator, and manipulation of the feedwater valve. A third operator may be similarly adjusting the de-aerator level and temperature (not shown in Figure 2-1).

The cost to engineer and install this system is low, but the labor cost associated with its operation is high, especially when one considers that most package boiler manufacturers offer automation packages requiring no operator attention to perform the above functions, while consistently maintaining better fuel efficiency and a more stable header pressure than the manual controls.

---

A manual control system is the most economical instrumentation system to purchase and install, however, the potential for uneconomical and unsafe operation is high due to the lack of equipment to monitor the process for human mistakes or equipment failure. The potential economical, safety, and environmental liabilities make manual control an undesirable option for most industrial processes.

## ❖ Chapter 3

# *Field Measurement Devices*

Instruments are available to continuously measure many process variables, the most common of which are flow, level, pressure, and temperature. Other process variables are routinely measured, including conductivity, density, pH, speed, position, current, voltage, and the like. Because a regulatory control loop uses only one device to measure the process variable, selection of the appropriate device to achieve *the desired measurement with acceptable accuracy at the proper location* is critical to the stable operation of the process.

In addition to evaluating a measurement device with respect to how well it performs, how long it maintains that performance, and how long it continues to operate without maintenance or failure must also be considered. Device failure and/or continued maintenance may be a symptom of a process problem, especially when the problem is related to the fluid part of the device.

While device replacement and/or upgrading might solve an immediate instrument problem, a more detailed analysis of the failure could lead to the correction of a process problem that, in turn, eliminates the instrument problem. It is not uncommon for properly designed measurement devices to operate reliably for years without failure.

To accompany the subtleties of measurement device selection, it is assumed that the reader has a working knowledge of these devices. Further insight into measurement may be obtained from the references listed at the end of this chapter.

It should be noted that selecting a device to measure a process variable may *appear* to be a relatively straightforward task, but, in practice, this may turn out to be quite a difficult exercise.

### Example 3-1:

Consider measuring the output pressure of a gas pressure regulator. It would appear that a pressure gage with an appropriate measurement range would suffice, if its construction materials were compatible with the fluid, because the downstream pressure will be stable at all times. However, it should be recognized that under no-load conditions, the gas pressure regulator may leak, causing the downstream measured pressure to eventually become equal to the upstream pressure. Therefore, the pressure gage must also be capable of withstanding the full upstream pressure without losing calibration or bursting.

Further, if the downstream pipe feeds a large vessel, the dynamics of the reaction in the vessel will affect the pressure downstream of the regulator, causing pressure fluctuations. Such dynamic conditions (on a seemingly static process) may prematurely wear out the mechanism of the pressure gage.

One problem associated with the selection of measurement devices is the assumption that the process is static. *Virtually all processes are dynamic and should be considered as such.* As illustrated above, often what *appears* to be an appropriate measurement technique may not be. Selection of an appropriate measurement device and options can be critical to achieving proper control.

## Flow Measurement

Although routinely applied to industrial processes, flowmeter installations are very likely to present measurement problems.

| Flowmeter Type | Principle of Measurement |
| --- | --- |
| Volumetric | Positive Displacement |
| Velocity | Magnetic |
| | Oscillatory |
| | Turbine |
| | Ultrasonic |
| Mass | Coriolis |
| | Hydraulic Wheatstone Bridge |
| | Thermal |
| Inferential | Differential Pressure |
| | Target |
| | Variable Area |

**Table 3-1.    Flow Measurement Methods**

It may seem that any flowmeter will perform adequately, how-ever, a more in-depth examination provides the insight that different flowmeters can measure flow very differently.

It must be understood that flowmeters are fundamentally me-chanical devices, even if they appear to be electronic, because the fluid passes through a flowmeter body. The volumetric fluid flow (Q) through a conduit (or flowmeter) with cross-sectional area (A) is related to fluid velocity (v) by

$$Q = A \times v \qquad \qquad (3\text{-}1)$$

In addition, the fluid mass flow (W) is related to the volumetric flow and fluid density by

$$W = \rho \times Q \qquad \qquad (3\text{-}2)$$

All flowmeters measure the fluid flow through the flowmeter body, but some flowmeters measure volumetric flow (Q) di-rectly, while others determine volumetric flow by measuring velocity (v). Mass flowmeters measure mass flow (W) directly, while other flowmeters are inferential in nature and measure a combination of mass, velocity, and density.

As shown in Table 3-1, there are four types of flowmeters, each of which measures flow differently (see Figs. 3-1 through 3-10). Selection of the appropriate flowmeter technology to perform the proper flow measurement for the application is of-ten the key to controlling the correct process variable.

### *Example 3-2:*

An understanding of the process is essential in the selection of the best flowmeter type for a given application.

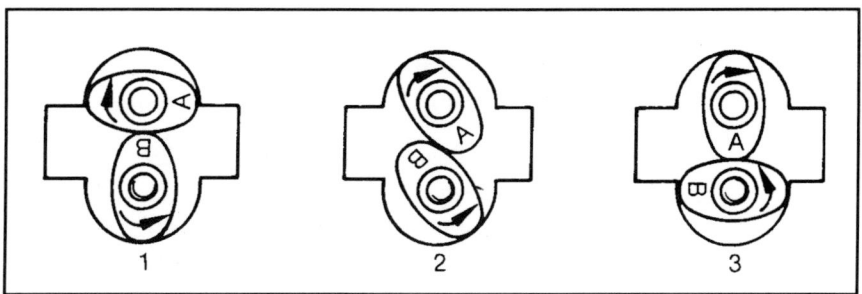

**FIG. 3-1.**    Oval gear positive displacement flowmeter (*Courtesy Brooks Instru-ment Division, Emerson Electric Company*)

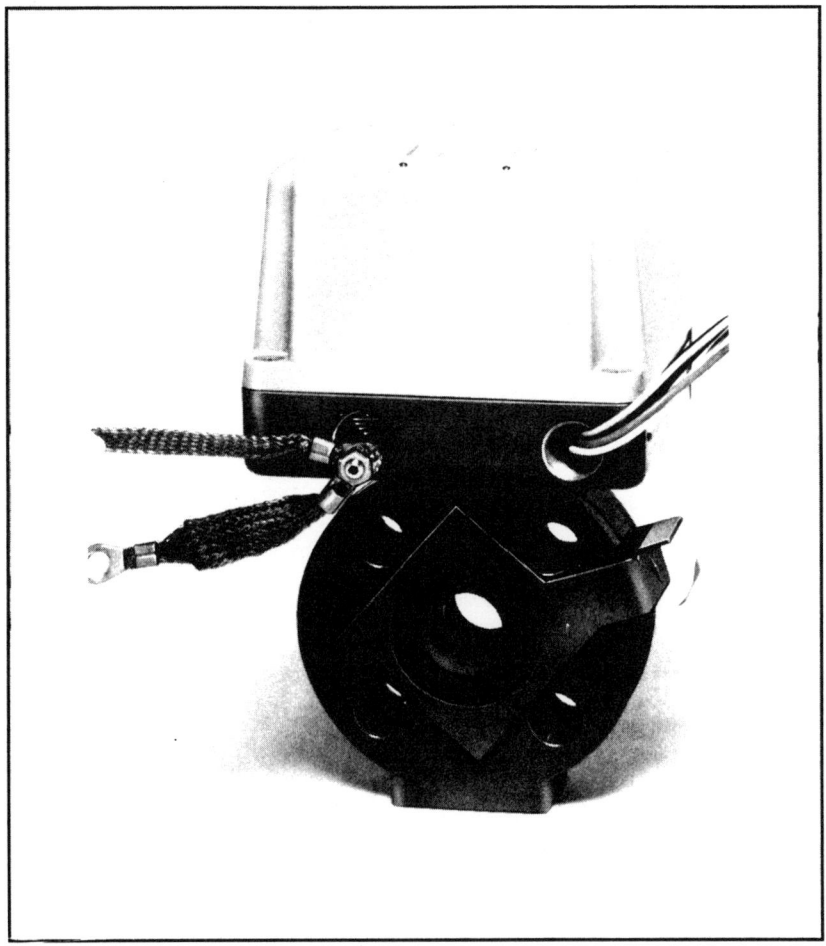

**FIG. 3-2.**  Magnetic flowmeter (*Courtesy Fischer & Porter Company*)

For example, raw materials feeding a chemical reactor may be measured best with mass flowmeters because chemical reactions are mass related. However, the high cost of a mass flowmeter may not be necessary for all of the raw materials if, for example, a raw material fed in excess, the reaction will be unaffected by the amount of excess, and the excess raw material will be recovered.

A flowmeter that measures volume or velocity may be applicable to filling a tank. If a tank truck is to be filled to a certain shipping weight, a mass flowmeter may be appropriate; however, if the operating fluid density does not vary appreciably, a volumetric or velocity flowmeter may exhibit acceptable performance.

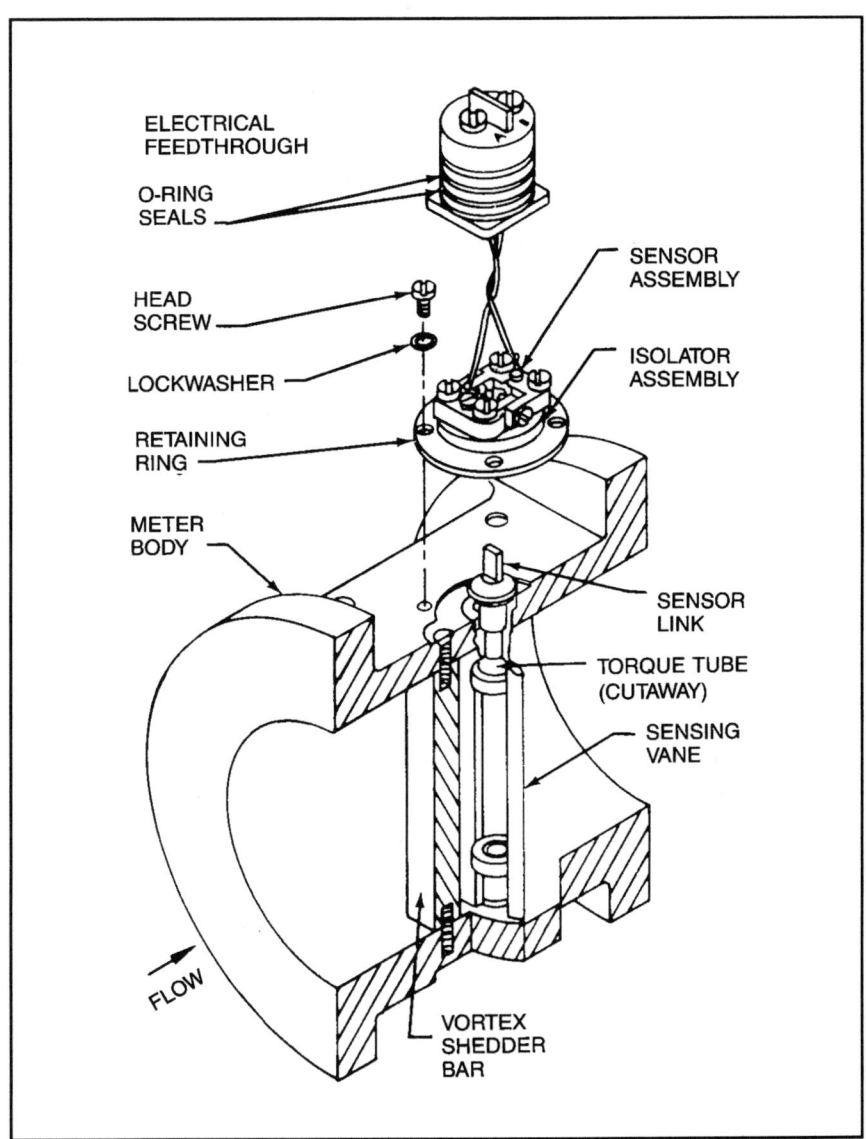

**FIG. 3-3.** Vortex shedding (oscillatory) flowmeter

In addition, most flowmeters, in order to achieve optimum accuracy, require that the fluid pass through the flowmeter body with a known velocity profile. This velocity profile is usually attained by: (1) installing the flowmeter to avoid two-phase flow (by keeping liquid flowmeters free of gas and gas flowmeters free of liquids); (2) providing an appropriate number of pipe diameters of straight run upstream and downstream of the flowmeter body; and (3) operating the flowmeter within a Reynolds number range in which the flowmeter performs

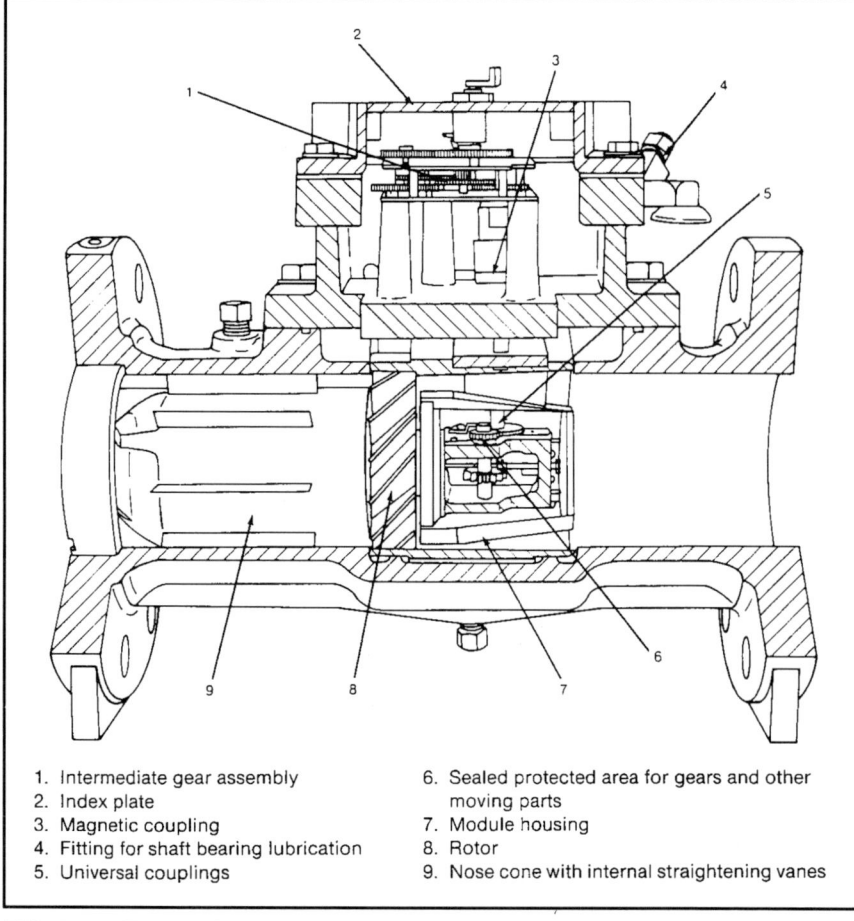

1. Intermediate gear assembly
2. Index plate
3. Magnetic coupling
4. Fitting for shaft bearing lubrication
5. Universal couplings
6. Sealed protected area for gears and other moving parts
7. Module housing
8. Rotor
9. Nose cone with internal straightening vanes

**FIG. 3-4.**   Turbine flowmeter

accurately. Straight run requirements vary with flowmeter technology and design; therefore, the use of rules of thumb should be avoided. It should be noted that improper installation (and lack of attention to detail) can easily cause a flowmeter to measure inaccurately.

Most flowmeters provide linear output signals; that is, the output signal is directly proportional to flow. Some flowmeters (most notably differential pressure flowmeters) have nonlinear output signals that vary with the square of flow.

It should be noted that the Coriolis mass flowmeter can be used to simultaneously measure the flow and the density of the fluid. Conceptually, the density measurement can be used to infer fluid composition by extrapolating the measured den-

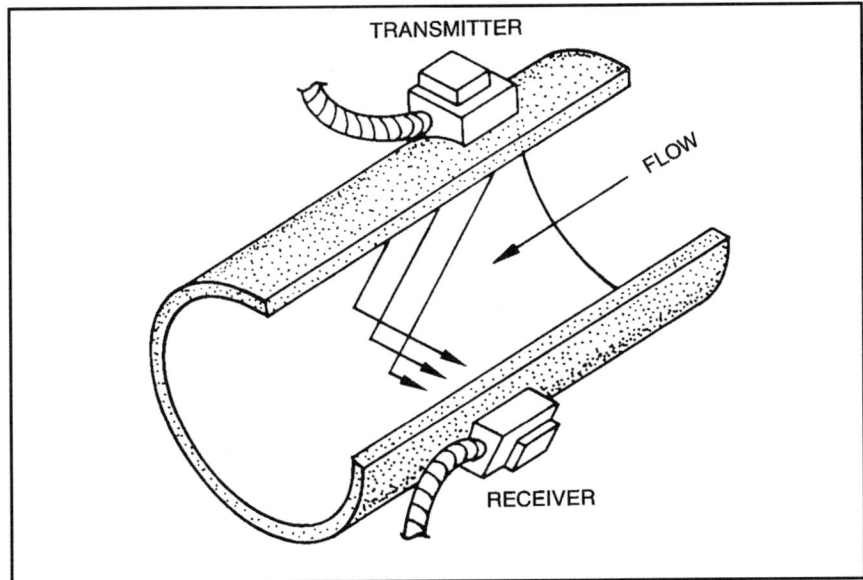

**FIG. 3-5.**   Ultrasonic flowmeter

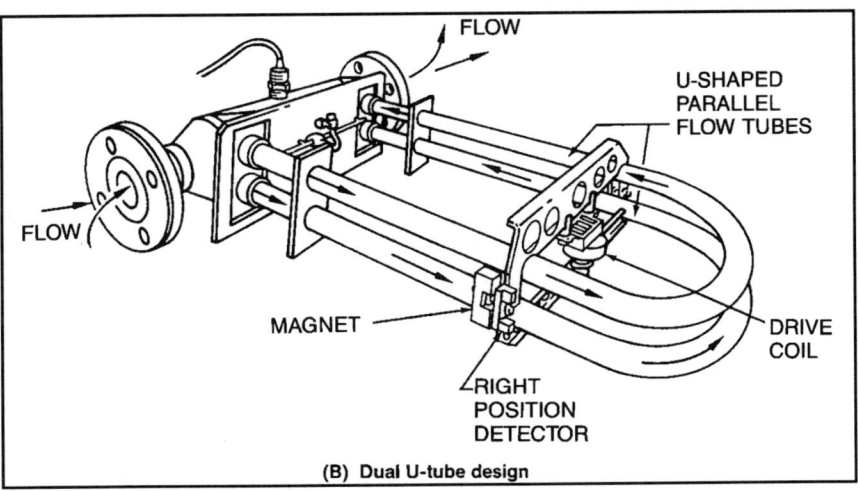

**FIG. 3-6.**   Coriolis mass flowmeter

sity between densities of the pure fluids. In practice, not only do the densities of the pure fluids vary with temperature, but the fluids may be miscible in varying degrees at different temperatures. Despite the complexity of the calculations, inferring fluid composition from the measured density can be an important process quality measurement.

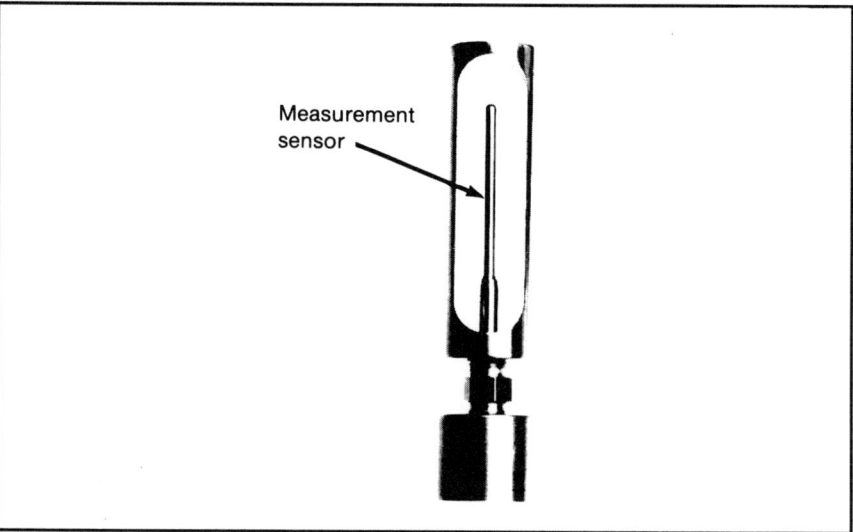

Measurement sensor

**FIG. 3-7.**  Thermal flowmeter (*Courtesy Kurz Instruments, Inc.*)

## Level Measurement

Level measurements are performed on vessels to control the inventory of the vessel. Most level measurement devices fall into one of three categories: those that measure level, those that infer level, and those that measure interface level.

Level measuring devices (see Figures. 3-11 through 3-16) that measure level directly include those that sense where the change of phase occurs in the vessel. The techniques used include floats, ultrasonics, radar, resistive tapes, capacitance, displacers, nuclear, and the like. Not all devices are applicable to a given application because they may be dependent upon the specific properties and characteristics of the process fluid.

Differential pressure level measurements detect the liquid head between a tap near the bottom of the tank and the gas space above the tank contents. This measurement does not sense at the location of the phase change, but rather is used in conjunction with the fluid density to infer the level in a vessel. Inaccuracies can occur because of temperature, pressure, and/or composition changes that cause variations in the fluid density.

Interface level measurements are usually more difficult to achieve than the level measurements described above. Typically, a differential pressure transmitter or a submerged displacer is installed across the interface to sense the overall

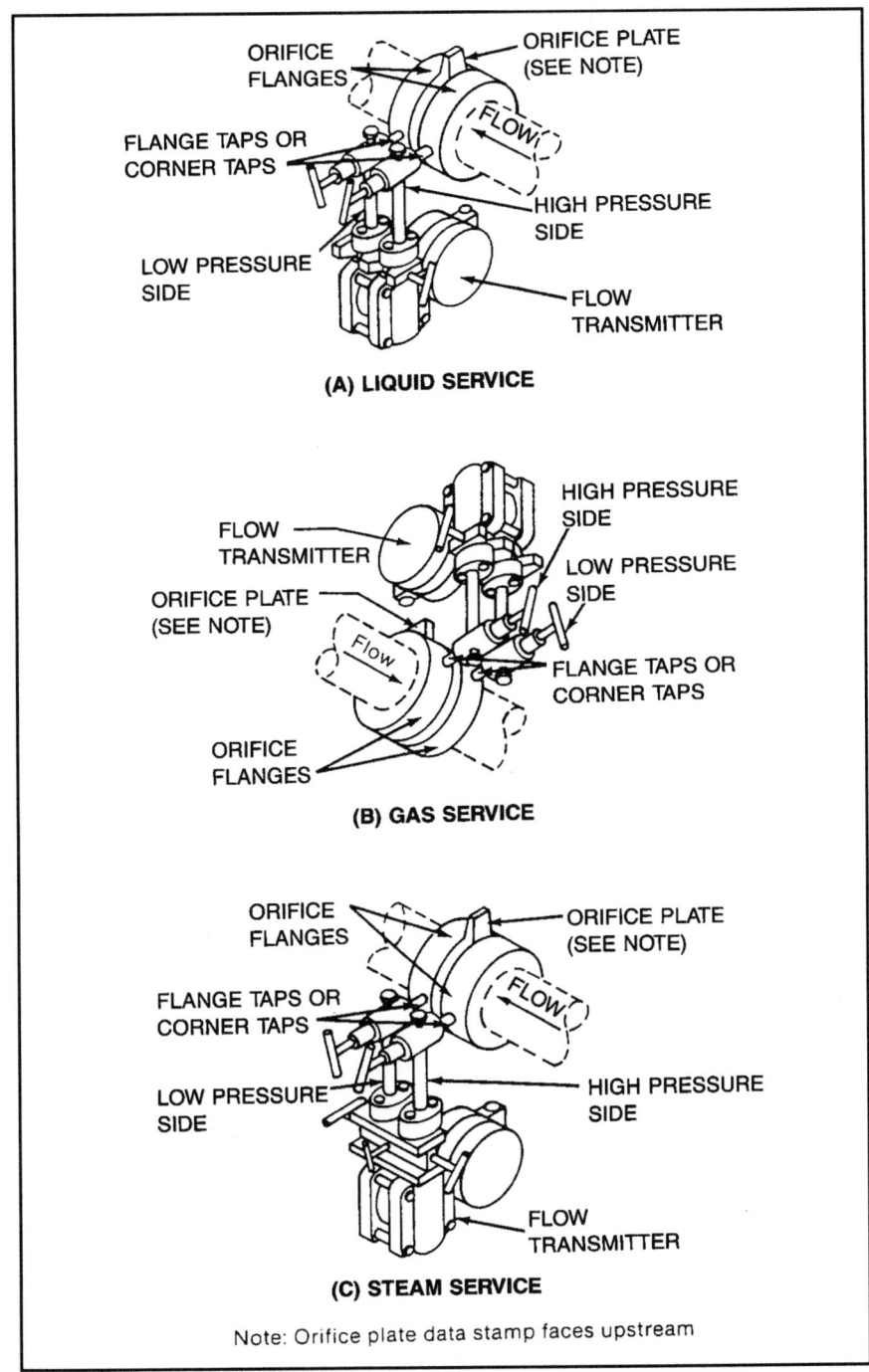

**(A) LIQUID SERVICE**

**(B) GAS SERVICE**

**(C) STEAM SERVICE**

Note: Orifice plate data stamp faces upstream

**FIG. 3-8.** Orifice plate (differential pressure) flowmeter

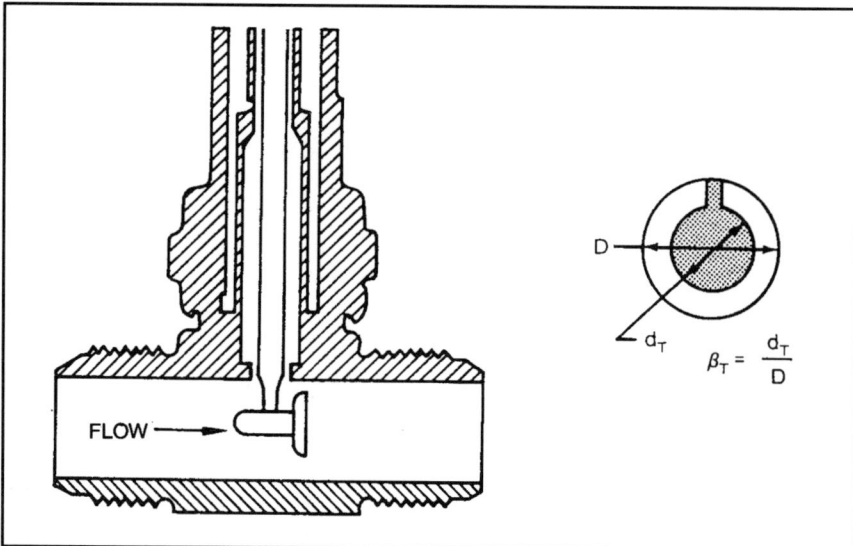

$$\beta_T = \frac{d_T}{D}$$

**FIG. 3-9.**    Target flowmeter (*Courtesy The Foxboro Company*)

density difference that occurs as the interface level changes. Capacitance technology can be used, if the liquids at the interface have sufficiently different electrical properties.

### *Example 3-3:*

When the fluid density does not vary appreciably, a differential pressure level measurement will suffice for measuring the liquid level in a storage tank that is dedicated to the fluid. However, if the tank were to be used for different fluids that have different densities, the differential pressure level measurement would be inaccurate, and a technique that measures level would be applicable.

## Pressure Measurement

Pressure transmitters are typically installed on pipes and equipment to measure pressure and to infer conditions inside the equipment (see Figures. 3-17). These transmitters are extremely reliable when properly installed and not abused (by steaming, for example); however, the pressure tap can plug and the transmitter can be affected by corrosion. In a given application, suitable materials of construction, the use of diaphragm seals, and proper installation can eliminate many of these problems. Although pressure instrument failure and/or

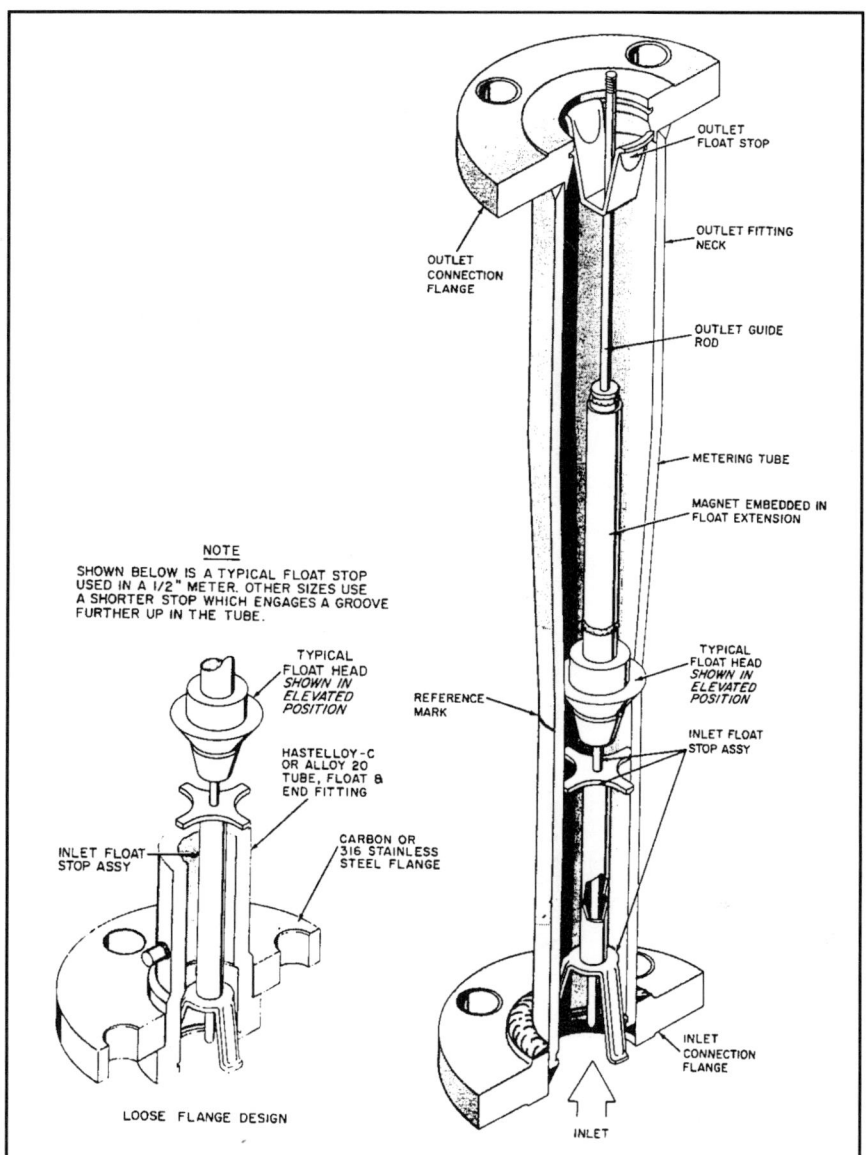

**FIG. 3-10.** Rotameter (variable area) flowmeter (*Courtesy Fischer & Porter Company*)

continued maintenance can be an indication of a process problem (which should be investigated), the problem usually lies with instrument installation or selection.

### Example 3-4:

Typical of gas service, the pressure transmitter is located above the pressure tap and valve that are installed on the side

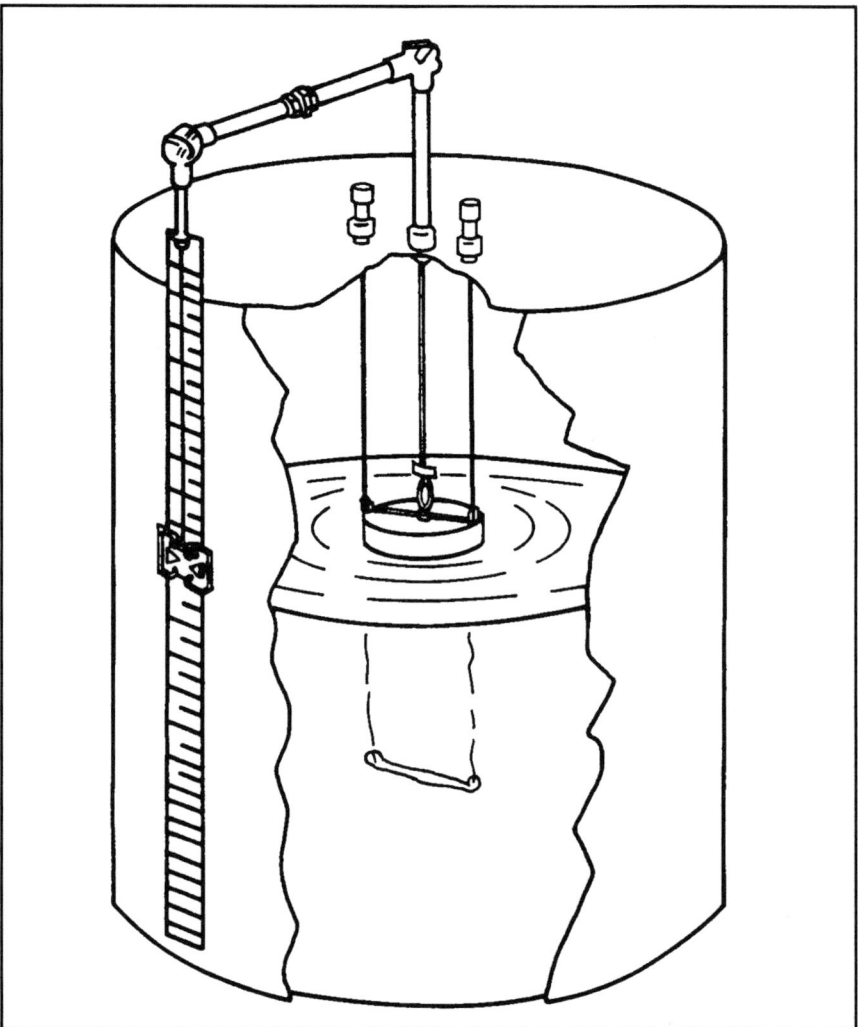

**FIG. 3-11.**    Float level transmitter

of a steam pipe. This installation will damage the transmitter because it allows live steam to enter and overheat the instrument. For hot condensable service, the transmitter should be installed below the tap (similar to a liquid application) so that the condensate forms a liquid seal that isolates the live steam from the transmitter.

## Temperature Measurement

A number of techniques are used to measure the temperature of industrial processes. These include many types of ther-

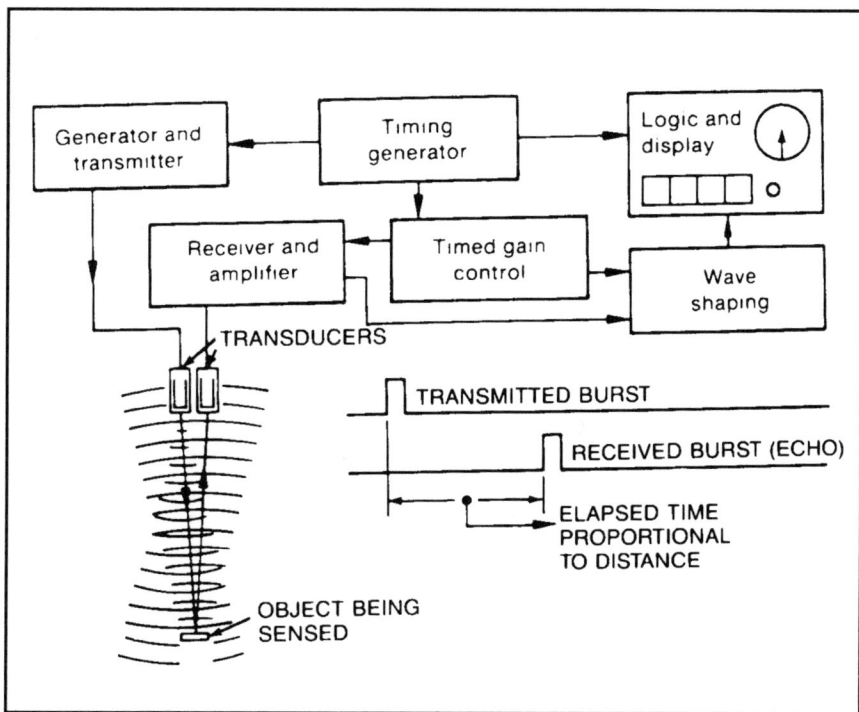

**FIG. 3-12.** Ultrasonic level transmitter (*Courtesy Gulf Publishing Company*)

mometers, thermocouples, thermopiles, thermistors, solid-state sensors, resistance temperature detectors (RTD), and infrared pyrometers (see Figures 3-18 through 3-20).

A number of temperature scales are in common use and misunderstandings can occur when temperatures are expressed in degrees without reference to the temperature scale being used. In process applications, misunderstandings most often occur between temperatures expressed as degrees Fahrenheit and degrees Celsius, however, absolute temperatures in the Rankine and Kelvin scales can also cause confusion.

It should be noted that, due to conduction, convection, and, occasionally, radiation effects, accurate industrial temperature measurement is difficult to achieve. Loss of measurement accuracy results from a lack of attention to detail and the difficulty in effectively compensating for these effects. In practice, these effects are often ignored entirely.

Noncontact temperature measurement and fast response time can be achieved by utilizing infrared pyrometer measurement

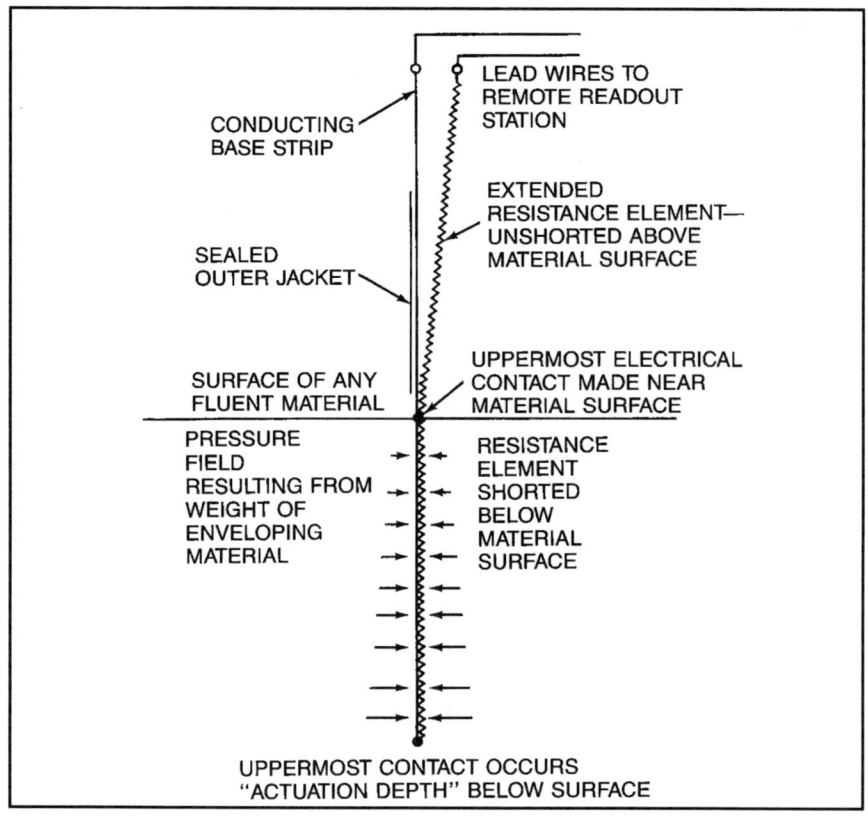

**FIG. 3-13.**    Resistive tape level transmitter (*Courtesy Metritape, Inc.*)

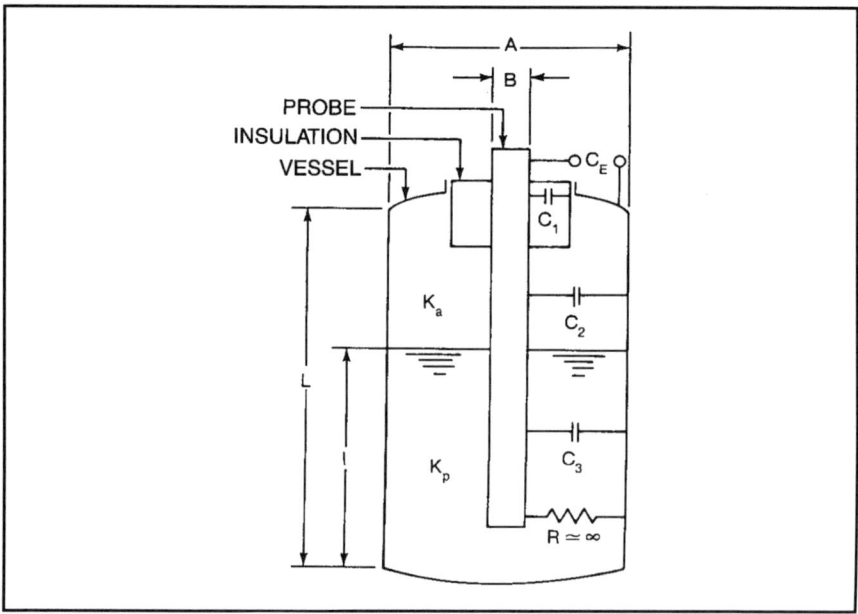

**FIG. 3-14(A).**    Capacitance level transmitter—nonconductive fluid

technology. Measurement inaccuracy can be introduced when the emissivity of the measured object is not accurately known.

### *Example 3-5:*

It is desired to measure the temperature of the flowing fluid in a pipe. A thermowell can be inserted into the pipe, and a thermocouple or RTD inserted into the well to measure fluid temperature. The mass of the thermowell and the thermal contact between the inner wall of the thermowell and the sensor will limit transient temperature response. When improved response is necessary, a spring-loaded sensor design or a thermowell with a built-in thermocouple design may be applied to improve this response, albeit at higher cost.

## "Smart" Transmitters

"Smart" transmitters that utilize microprocessor electronics are more stable and can usually provide more accurate data than their analog counterparts because of the inherent stability of digital circuitry and cost-effectiveness of implementing more sophisticated algorithms. One salient advantage of this technology is the ability to accurately measure the process variable over small ranges.

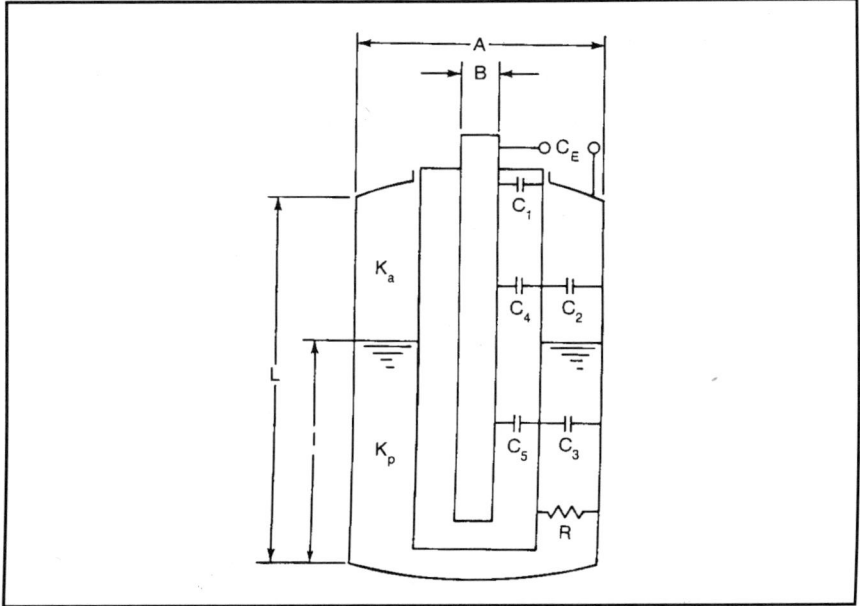

**FIG. 3-14(B).**   Capacitance level transmitter—conductive fluid

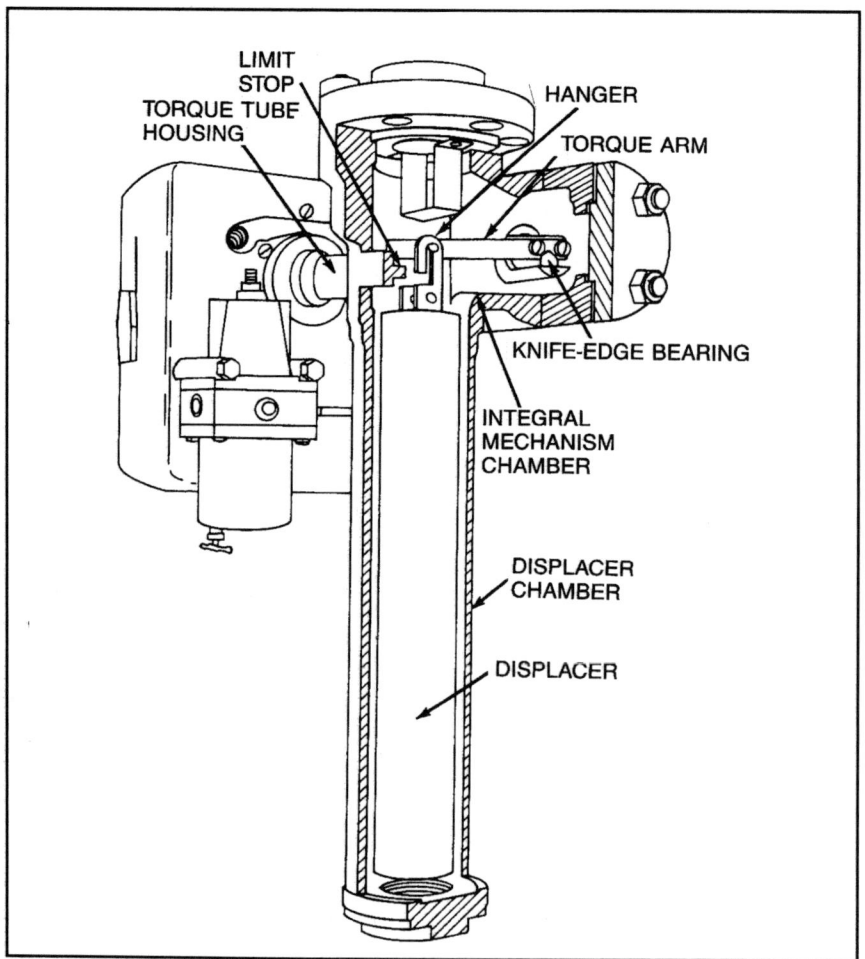

**FIG. 3-15.**    Displacer level transmitter (*Courtesy Masoneilan Division of McGraw-Edison Co.*)

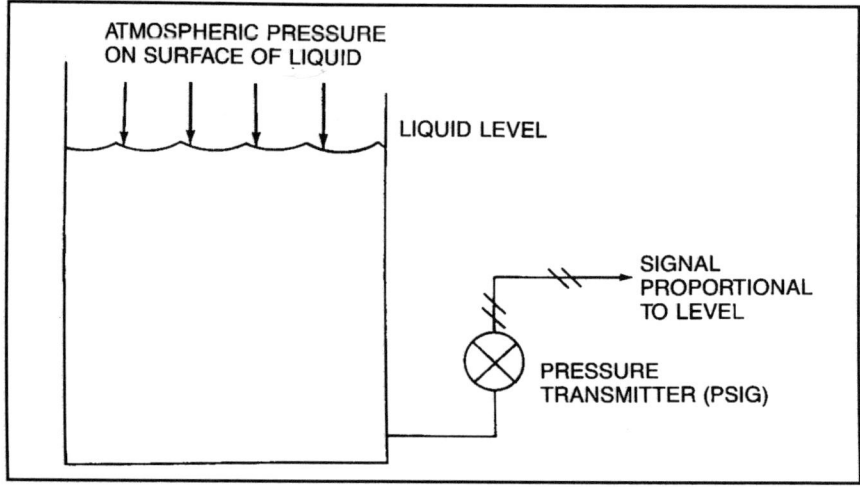

**FIG. 3-16(A).**    Differential pressure transmitter—open tank

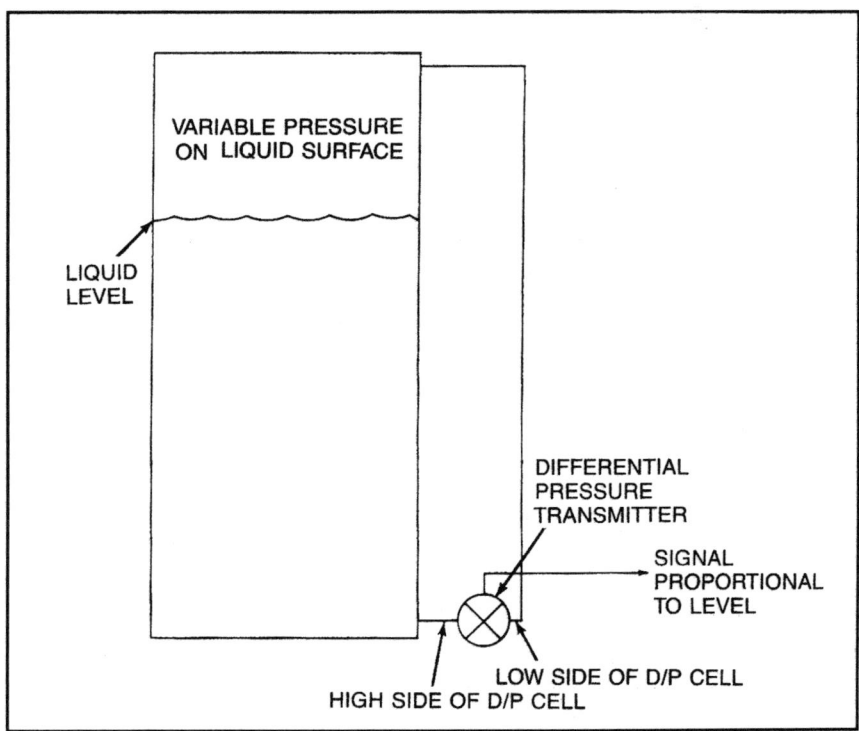

**FIG. 3-16(B).** Differential pressure transmitter—closed tank

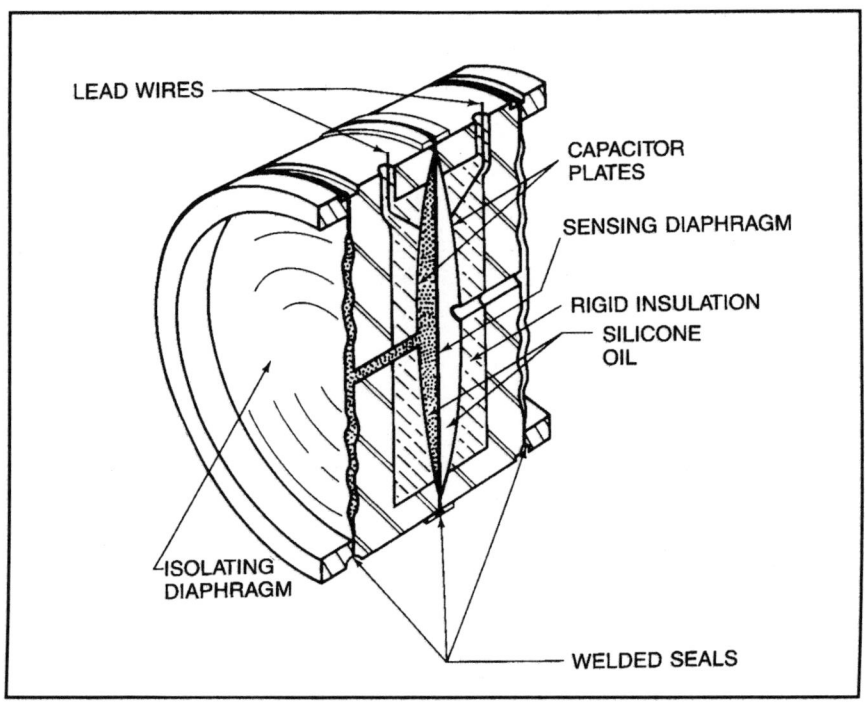

**FIG. 3-17.** Pressure transmitter sensing module (*Courtesy Rosemount Co.*)

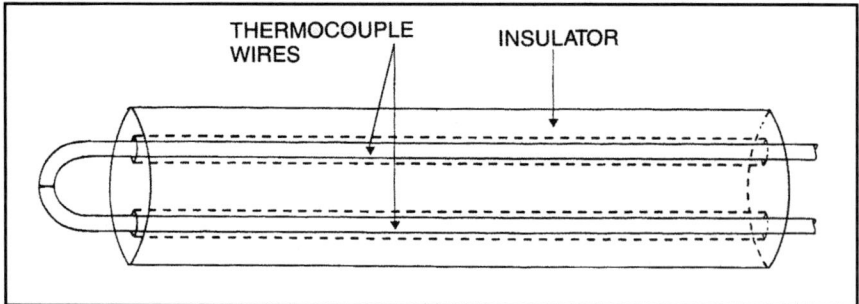

**FIG. 3-18.**   Thermocouple

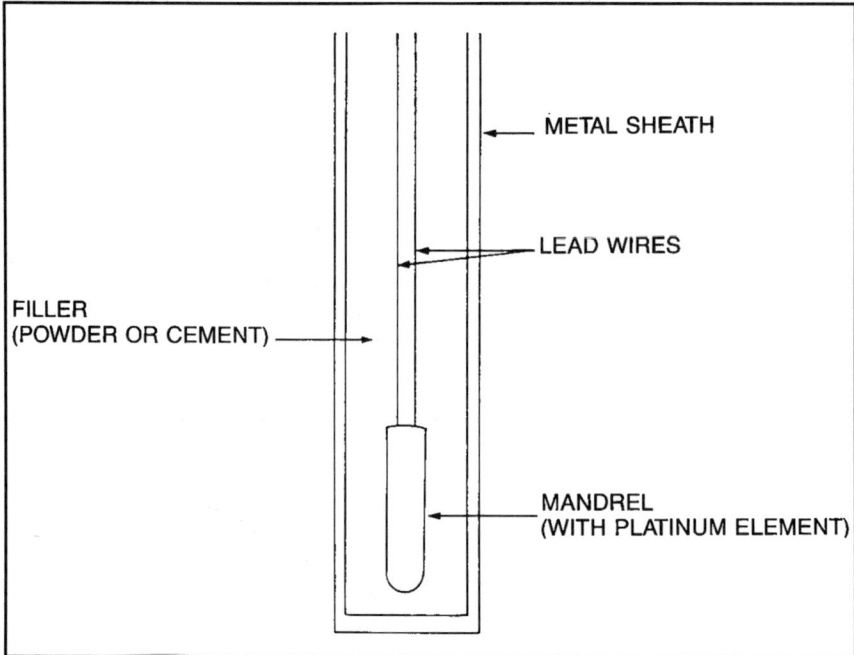

**FIG. 3-19.**   Resistance temperature detector

In addition to the digital transmission of process information, these "smart" transmitters can provide additional information to the user through a hand-held configurator or a distributed control system interface (such as transmitter diagnostics to aid in troubleshooting and to provide maintenance information) and can be electronically re-ranged.

## Summary

The quality with which a measurement is made in the context of an application represents an extremely important aspect of

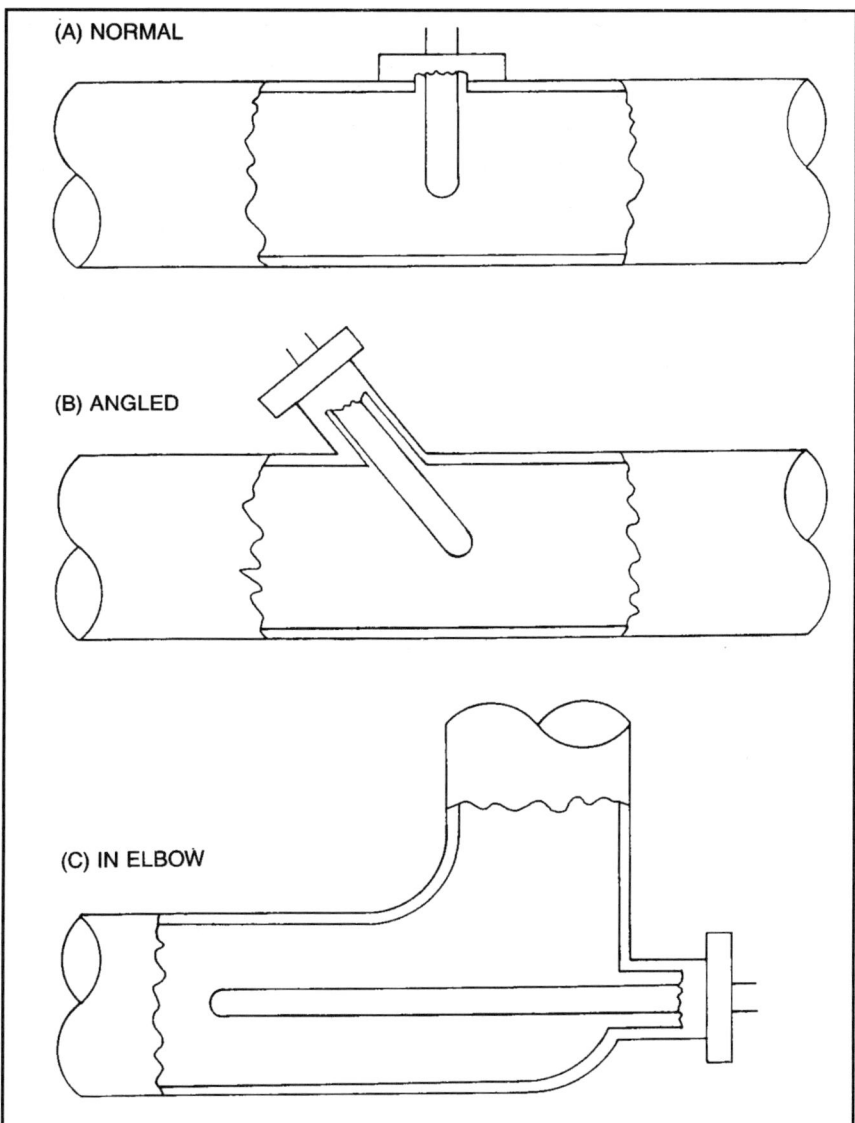

**FIG. 3-20.** Thermowell installations

control loop design. Not only must the measurement be the correct measurement for the application, but it must also be a faithful representation of the desired part of the process. If this is not the case, a control loop may appear to be functioning properly while the actual process is out of control.

## For Further Information

1. DeCarlo, Joseph P. 1984. *Fundamentals of Flow Measurement.* Research Triangle Park, NC: ISA.

2. Gillum, Donald R. 1982. *Industrial Pressure Measurement.* Research Triangle Park, NC: ISA.
3. Gillum, Donald R. 1984. *Industrial Level Measurement.* Re search Triangle Park, NC: ISA.
4. Kerlin, Thomas W., Shepard, Robert L. 1982. *Industrial Temperature Measurement.* Research Triangle Park, NC: ISA.
5. Miller, Richard W. 1989. *Flow Measurement Engineering Handbook;* McGraw- Hill.
6. Spitzer, David William. 1990. *Industrial Flow Measurement.* Research Triangle Park, NC: ISA.
7. Spitzer, David William. 1992. *Flow Measurement.* Research Triangle Park, NC: ISA.

# Chapter 4

# *Controllers*

Regulatory control is used extensively throughout industry to control continuous processes. Properly applying this technology can yield control that is vastly superior to manual control. Improved control can significantly improve yield, increase capacity, increase plant safety, and reduce the number of operators required to safely operate the plant. The centralization of process monitoring and process adjustments requires fewer operators and allows those operators to control the plant more safely without being physically located in the processing area.

The implementation of a regulatory control system requires considerably more control system engineering expertise than is required for a manual system. The cost of implementation is also considerably higher because better quality instrumentation is used, additional wiring and piping are necessary, and a control room must be constructed.

Figure 4-1 illustrates the components of a basic regulatory control loop. A measurement device senses the process and generates a signal that is transmitted to the controller. The controller compares the measurement with the desired set point that has been set by the operator; its output manipulates the final control element, typically a valve. *A number of disturbances can upset the process* and cause the measurement to move away from its set point. *The quality of a control system is, among other aspects, determined by how well the system compensates for these disturbances* to safely maintain the process at its desired operating condition.

### *Example 4-1:*

A regulatory control loop in common use is the cruise control system installed on some cars (Figure 4-2). The vehicle speed is sensed by the speedometer (measurement device), which feeds the speed measurement to the speed controller. When the desired speed is attained, the operator engages the speed

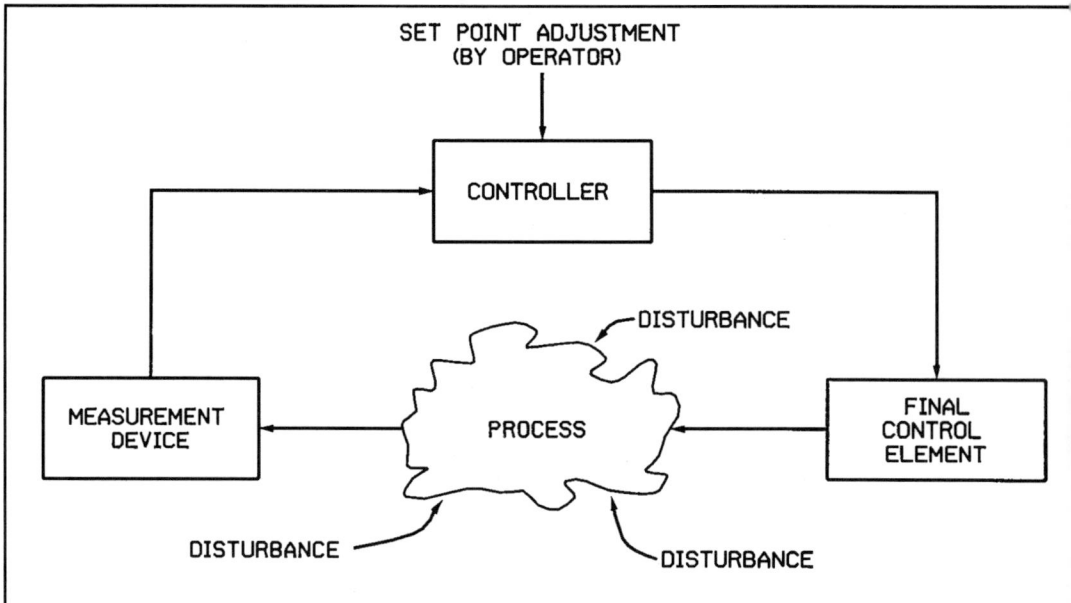

**FIG. 4-1.**    Components of a basic regulatory control loop

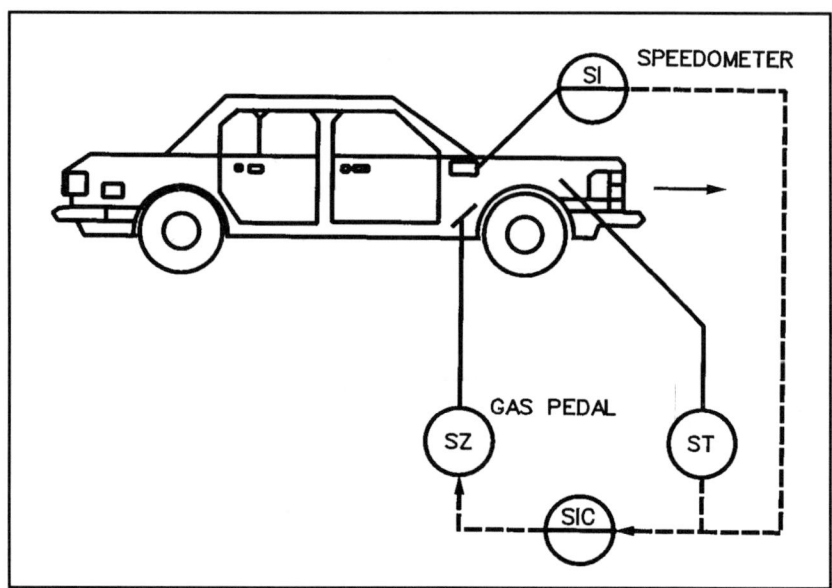

**FIG. 4-2.**    Cruise control loop

control, which also sets the speed set point. The speed controller then attempts to maintain the vehicle at constant speed by increasing the depression of the gas pedal (final control element) when the speed decreases and decreasing the depression of the gas pedal when speed increases. Note that, while moving the gas pedal, the final control element manipulates fuel flow.

This would seem to be a straightforward control system until the effects of disturbances are considered. What if the road becomes slippery and control of the car is lost because the speed control accelerated too much? What if the driver must stop suddenly, but the driver's foot is resting in such a position that it is difficult to move to the brake pedal quickly? Such disturbances (although having a small probability of occurring) are not handled well by this control system, and another control loop (anti-lock brakes, for example) may be required.

## Controlling Instruments

Controlling part of a process entails maintaining a certain measured parameter called the process variable (PV) at its desired value or set point (SP). This is achieved by manipulating the process in such a manner that the process variable reaches the set point. A controller is a device that compares the process variable to the set point and generates an output signal (OUT) that manipulates the process in order to make the process variable equal to its set point. Mechanical, pneumatic, electronic, and microprocessor-based controllers are available, each potentially providing progressively more sophisticated performance.

Most controllers display the set point, the process variable, and the output; they can be changed from the manual (MAN) to the automatic (AUTO) mode and vice versa by the operator. In the manual mode, the operator can directly adjust the controller output, but set point changes made by the operator have no effect on the controller output. When in the automatic mode, the controller compares the process variable to the set point and utilizes its control algorithms to generate the controller output, but output changes made by the operator have no effect on the controller output.

Most controllers commonly used in the process industries contain the features described above (see Figure 4-3) and use knobs and buttons to change modes and adjust the controller set point and output. Certain lower-cost mechanical and pneumatic controllers may or may not contain all of the above features.

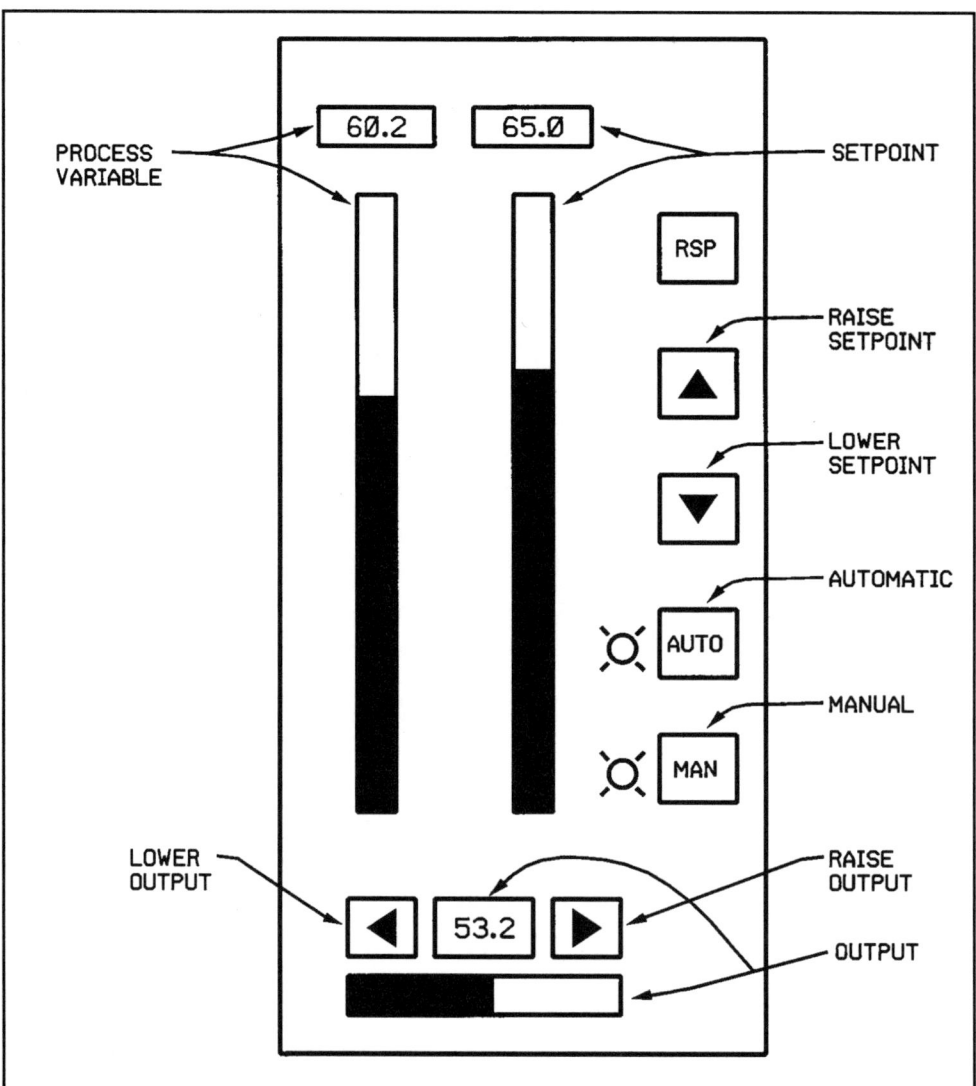

**FIG. 4-3.**   Typical microprocessor-based controller faceplate

## Local Control

A local regulatory control loop, as illustrated in Figure 4-1, is located in the field and may indicate and/or record locally. These loops typically consist of one or more pieces of equipment; the number depends upon whether the instrument is designed to combine functions, such as in the case of a mechanical float valve assembly that measures, controls, and manipulates the control valve using essentially one piece of equipment.

Loop sophistication (that is, the decision to use mechanical, pneumatic, and/or electronic devices) is typically dependent upon the available power source, the desired control loop performance, and cost.

### Example 4-2:

Figs. 4-4, 4-5, and 4-6 illustrate progressively more sophisticated local, tank temperature control loops. The mechanical control loop (Figure 4-4) is a self-contained regulating valve that requires no external power source. Expansion of the fill fluid in the temperature probe opposes a spring (set by the operator) and mechanically operates the steam valve (manipulating steam flow), thereby controlling tank temperature. Rudimentary mechanical adjustments may be available to "tune" the loop, but these adjustments are usually made at the factory and are not touched in the field.

Superior control can be obtained by using the system shown in Figure 4-5, in which the expansion of the fill fluid in the temperature probe is compared to the set point, generating an output signal to the control valve. Mechanical adjustments can be made to the controller to improve system performance.

Further control improvement can be obtained by using the loop illustrated in Figure 4-6, in which electronic devices

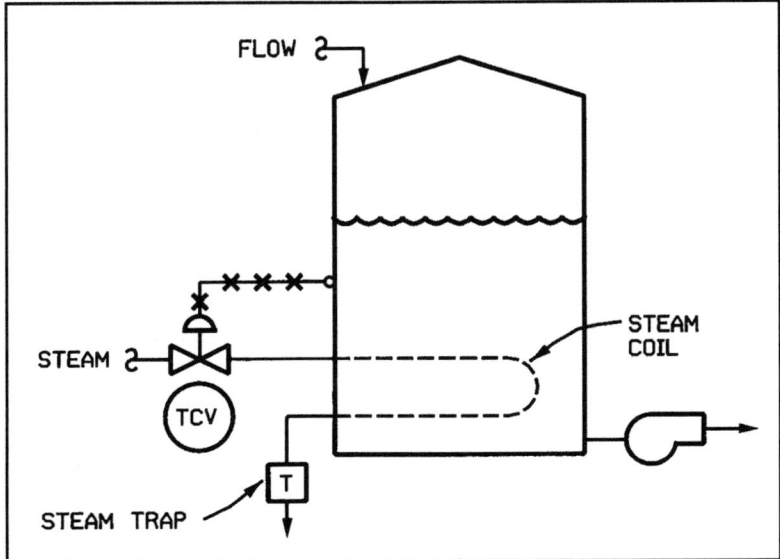

**FIG. 4-4.** Local regulatory control loop: tank temperature control with mechanical temperature control valve

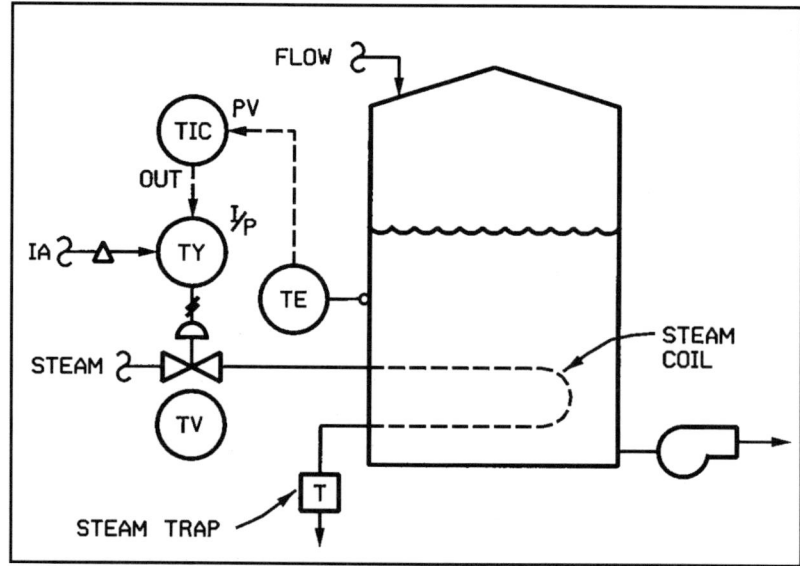

**FIG. 4-5.**   Local regulatory control loop: tank temperature control with pneumatic controller and control valve

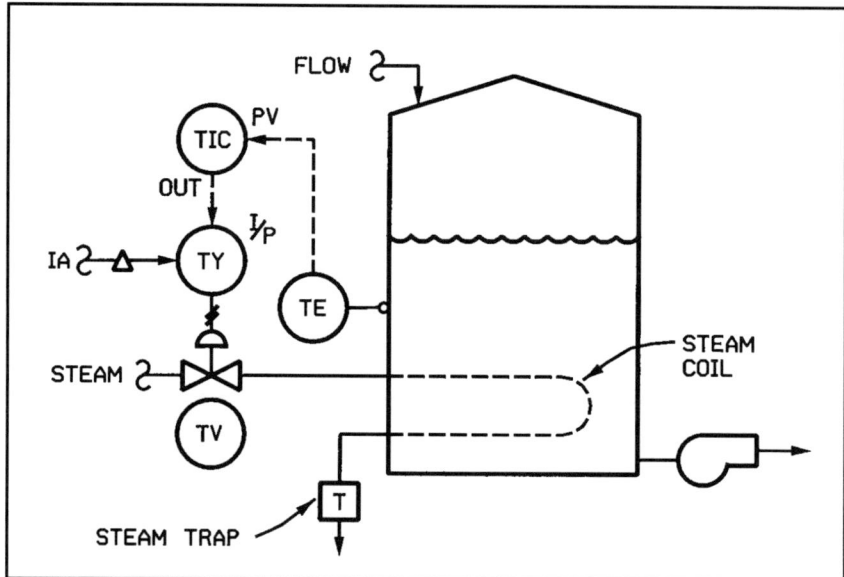

**FIG. 4-6.**   Local regulatory control loop: tank temperature control with electronic controller and pneumatic control valve

are used for their improved accuracy, speed, stability, and flexibility. A wider range of controller adjustments than those available on local pneumatic controllers can used to further improve system performance.

# Centralized Control

A centralized regulatory control loop performs the functions illustrated in Figure 4-1, where the control loop consists of a measurement device, a controller, and a final control element. Signals from the measurement device and to the final control element are routed, using pneumatic or electronic technology, to and from a controller located in a central control room or on a centralized control panel. A recorder and/or alarm annunciator are often mounted on the control panel near the controller to record and/or alarm the process variable.

***Analog Signal Transmission*** Pneumatic or electronic techniques can be used to represent the magnitude of an instrument signal in a continuous manner. The 3-15 psig (0.2-1.0 bar) pneumatic signal that can be used to represent 0-100 percent signal is transmitted from one location to another using pneumatic tubing. Pneumatic repeaters may be necessary to minimize the time required to transmit a signal through long tubing runs. Electronic signals, such as the 4-20 mA signal that can represent 0-100 percent signal, provide virtually instantaneous signal transmission.

***Digital Signal Transmission*** In a regulatory control loop, digital signal transmission is functionally equivalent to analog signal transmission (described above) except that the controller receives its measurement data and/or sends its output signal as a digital signal; that is, the controller communicates with the transmitter and/or final control element digitally where the controller and the communicating device(s) are designed (and configured) to communicate using a common protocol. Digital signal transmission using "smart" transmitters can improve measurement accuracy and diagnostics and can provide additional maintenance features. Controller digital communications can be implemented using an appropriate controller or a distributed control system with appropriate modules.

### Example 4-3:

Figure 4-7 illustrates a pneumatic regulatory level control loop in which the level transmitter and control valve are located in the field and the controller is located in the control room. In this loop, the level transmitter transmits a 3-15 psig (0.2-1.0 bar) signal, proportional to the tank level, to the controller

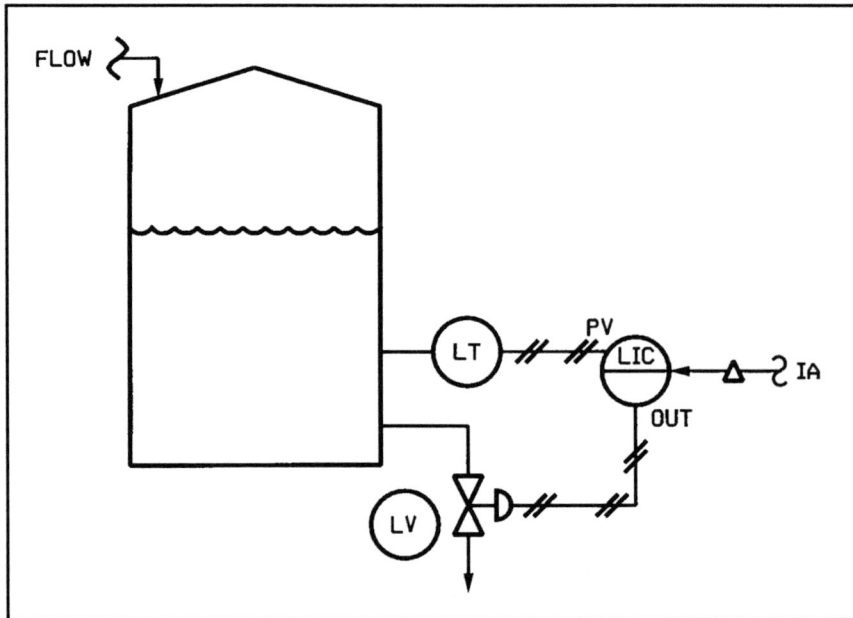

**FIG. 4-7.**   Centralized analog regulatory control: pneumatic level control loop

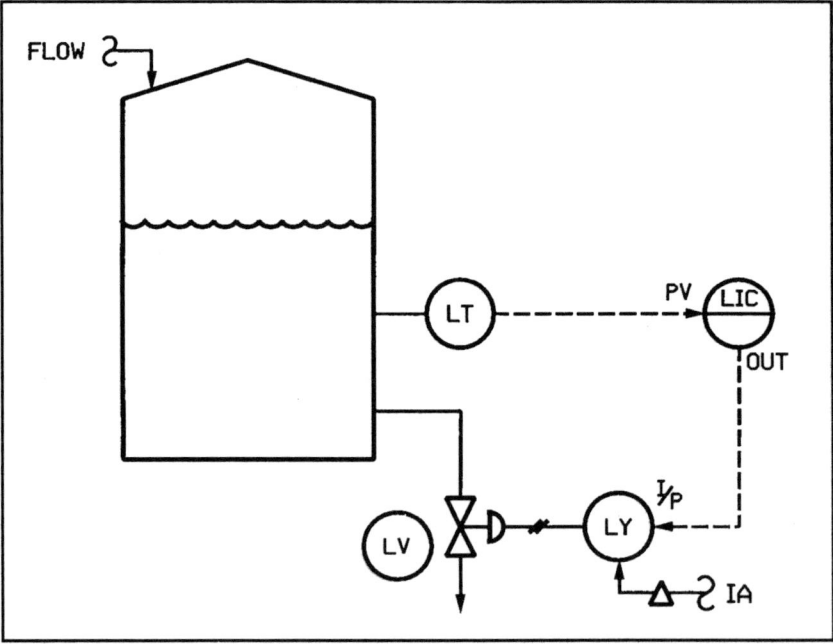

**FIG. 4-8.**   Centralized analog regulatory control: electronic level control loop

process variable input. The controller operates the control valve (manipulating flow) by transmitting a 3-15 psig (0.2-1.0 bar) signal to the control valve actuator.

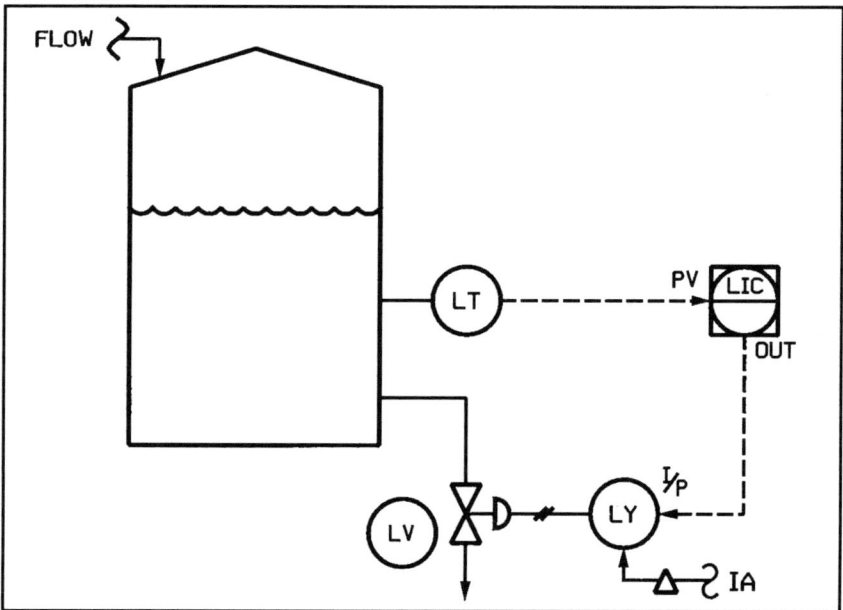

**FIG. 4-9.**   Distributed control system regulatory control: electronic level control loop

Figure 4-8 illustrates the same loop augmented by electronic instrumentation (which may use analog or digital electronics) communicating through 4-20 mA signals or digital protocol. The electronic signal is converted to a pneumatic signal in the field to operate the valve.

## Distributed Control

The implementation of regulatory control using a distributed control system is functionally equivalent to the centralized control described above; however, the system is designed in such a manner that the measurement input, the controller software, and the control valve output may be in different plant locations. Controller information can be made available on the display(s) located in the control room and (potentially) throughout the plant. (Figure 4-9)

## Summary

Understanding the fundamental qualitative operation of controllers and the methods by which measurement signals are transmitted to the controller are important. These concepts are especially important for locating a problem when troubleshooting the control loop.

# Chapter 5

# Final Control Elements

Final control elements are those devices that manipulate the process. The most common final control element applied to industrial processes is the control valve; however, in some applications, a number of final control element technologies can be used. These alternate technologies, primarily variable speed drives, can often provide performance and/or features that are superior to those of control valves.

*IT SHOULD BE CLEARLY NOTED THAT VIRTUALLY ALL FINAL CONTROL ELEMENTS MANIPULATE FLOW.* Varying the opening of a control valve, the opening of a damper, or the speed of a variable speed drive alters process flow. Heating (cooling) is effected by manipulating the flow of the heat (cooling) source, such as steam, hot oil, fuel, electricity, cooling water, brine, and the like.

This chapter discusses the importance of the final control element as it relates to the control loop. Further insight into the operation and application of final control elements may be obtained from the references.

## Control Valves

Control valves are devices that dissipate hydraulic energy in a controlled manner to achieve a desired flow. The flow through a piping system without a control valve will be determined by where the pump curve (which defines pump performance) intersects the system curve (which defines piping head and friction losses). See Figure 5-1. The flow through the system can be varied by throttling a control valve, effectively increasing the piping friction. See Figure 5-2. In this manner, flow can be manipulated from a minimum of no flow to the maximum flow produced when the control valve is in its fully open position. Note that a fully open control valve still throttles the piping system.

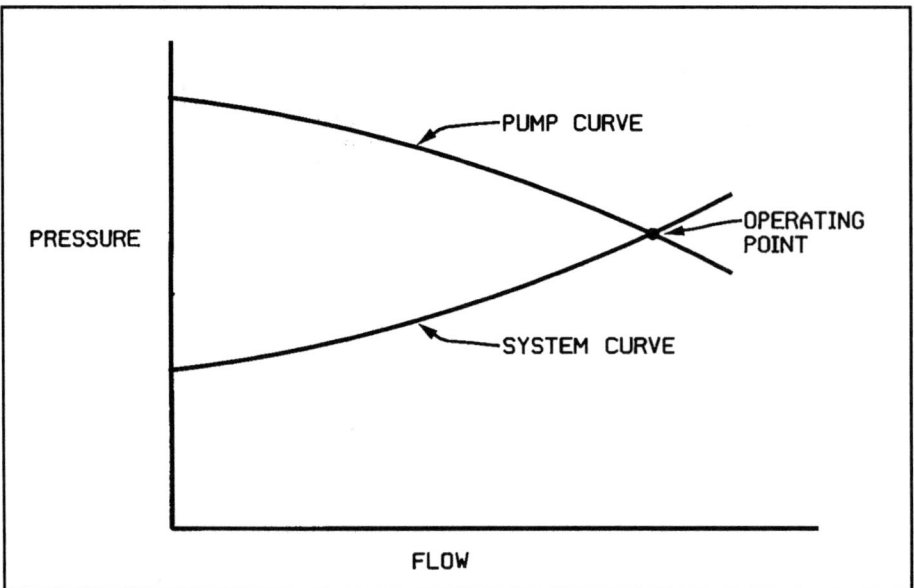

**FIG. 5-1.**   Operating point of unthrottled piping system

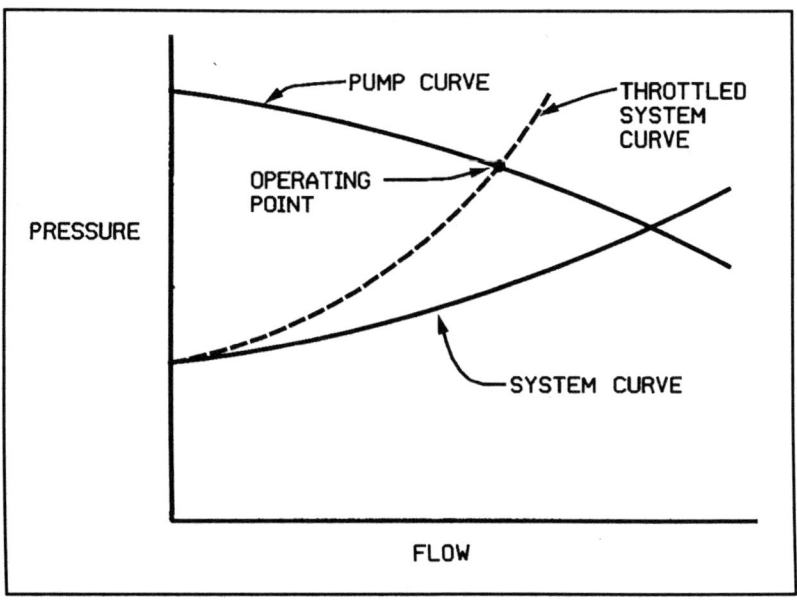

**FIG. 5-2.**   Operating point of throttled piping system

Control valve selection and sizing constitute a challenging process involving many considerations. The control valve must be sized to pass the maximum (design) flow at a low enough pressure drop to require no more pressure than the pump is capable of delivering, yet maintain a high enough pressure drop to effect good control. Undersized valves will not pass enough flow, which presents an immediate problem. Oversized valves usually operate nearly closed resulting in poor control of flow because small valve position changes result in large changes in flow.

It is not uncommon to violate one of the above constraints, especially when using high-capacity valves, such as ball or butterfly valves, that are smaller than line size. On start-up, construction debris can lodge in a valve and make the valve appear to be undersized.

Another concern is the hysteresis associated with control valves, which can cause the valve not to operate smoothly. As a result of the tightening of valve fugitive emission regulations, valve leaks will increasingly be addressed by adjustment of the packing to minimize leakage. The problem occurs when the person adjusting the packing uses the philosophy that "if a little is good, more is better" and adjusts the packing extra tight so it is less likely to leak. What is not realized is that after the adjustment, the valve may operate sluggishly or not move at all.

Control valves are typically remotely operated through pneumatic diaphragm actuators. A current-to-pneumatic (I/P) converter is used to linearly convert the electronic signal to a pneumatic signal that is applied to the actuator. The electronic signal can be varied, which in turn varies the pneumatic signal, which in turn strokes the valve.

It should be noted that the electronic signal varies the pneumatic signal to the control valve actuator. Valve travel is in response to the pneumatic signal, and *the relationship between the electronic signal and valve position is not necessarily linear*. Valve position is dependent upon the forces acting on the valve and the actuator, and the valve may exhibit hysteresis, where the same signal may result in different control valve positions, depending upon the direction of travel, fluid forces, and friction. Further, an undersized valve actuator can cause control problems, such as when the valve cannot be opened far enough to provide the desired flow.

Valve positioners are installed on some control valves to minimize the hysteresis and the valve position inaccuracy described above. These devices are self-contained feedback control loops that measure the actual valve position, detect the desired valve position, and provide the proper pneumatic actuator pressure to move the valve to the desired valve position. Thus, the electronic signal and the valve position are linearly related.

The relationship between valve position and valve capacity is dependent upon the valve trim characteristic (such as linear, equal percent, or quick opening). In addition, the relationship between valve position and the actual (installed) valve capacity depends upon the valve characteristic and the actual hydraulics of the piping system. Selection of a valve trim characteristic that results in an inappropriate installed valve charecteristic for a given piping system can cause control problems.

The control valve must move to its proper failure mode on loss of power or instrument air. Most valves in industrial applications fail closed (FC), however, many valves fail open (FO) and or fail-in-last-position (FLP) depending upon the application.

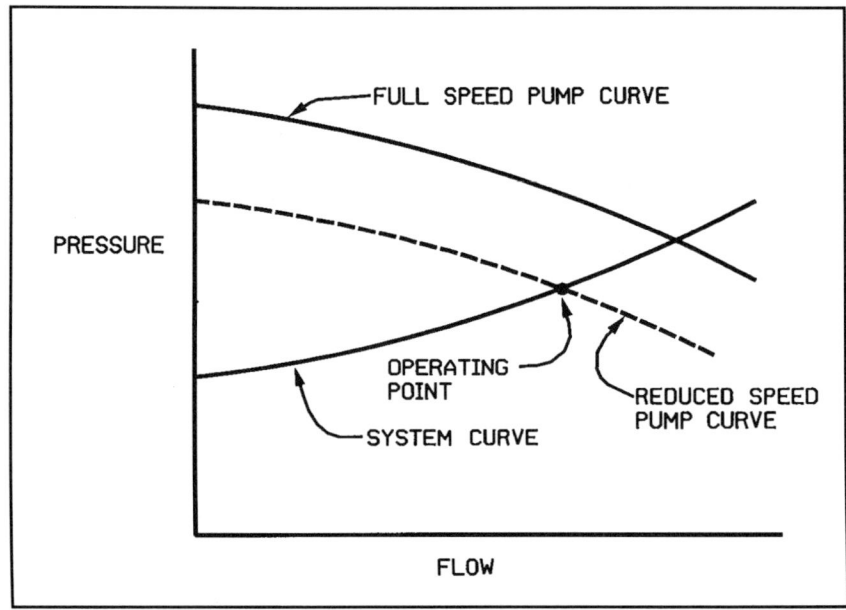

**FIG. 5-3.**  Operating point of piping system throttled with a variable speed drive

# Variable-Speed Drives

Most variable-speed drives vary motor speed, which in turn varies the speed of the load, however, other variable-speed drives vary only the speed of the load. Significant energy savings can be achieved using either technology to generate only the amount of hydraulic energy needed by the process; however, varying motor speed will typically yield larger savings and exhibit little of the hysteresis that can plague control valves. In many applications, the installed cost of variable-speed drives can be less than that of control valve installations.

Most applications are electronic, variable-frequency drives applied to centrifugal loads where energy consumption is related to the cube of motor speed. Therefore, operating the motor at 90 percent of full speed (a modest reduction) will reduce energy consumption by approximately 27 percent, illustrating the potential energy savings.

Variable-speed drives cannot provide the tight control valve shutoff capability of a control valve, but flow in the reverse direction can be minimized by installing a check valve downstream of the pump or fan. When tight shutoff is required, an on/off valve may be required.

Variable-speed drive technology is not applied to its full potential (to improve control, reduce the number of leak paths in the process, reduce maintenance costs, and provide significant energy savings) because: (1) control valves are applied routinely; (2) many instrument engineers do not understand the performance of hydraulic equipment; and (3) the instrument engineer typically does not have as much control over the application of this technology as compared to control valves.

### Example 5-1:

Control valves and variable-speed drives can be used to throttle flow. These systems perform the same function, that is, providing the desired flow; however, the variable-speed drive will generate only the hydraulic energy necessary to deliver the desired flow. The full-speed pump will generate excess hydraulic energy, which must be dissipated across the control valve. Varying the speed will be more energy efficient, eliminate leak paths, minimize hysteresis, and subject the pump and piping to lower pressures, which can reduce the potential for leaks. Compare Figures 5-2 and 5-3, and see Figure 5-4.

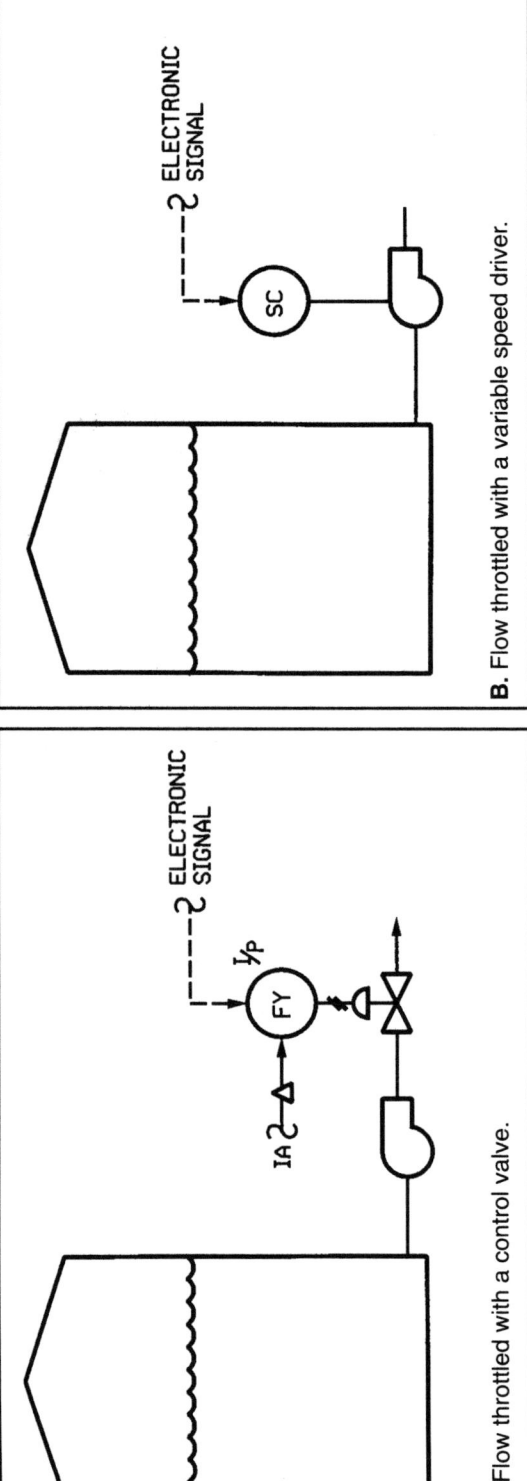

**A.** Flow throttled with a control valve.

**B.** Flow throttled with a variable speed driver.

**FIG. 5-4.** Throttling flow by different methods

# Summary

In many applications, lack of attention to detail during control valve selection can result in potential control problems, possible safety hazards, and excessive maintenance requirements, which may compromise the effectiveness of the control system.

# References

*Control Valve Handbook,* Fisher Controls, 1977
Hutchison, J. W. (editor). 1976. *ISA Handbook of Control Valves.* Research Triangle Park, NC: ISA.
Spitzer, David William. 1990. *Variable Speed Drives—Principles and Applications for Energy Cost Savings.* Research Triangle Park, NC: ISA.

# Chapter 6

# *Field Equipment Tuning*

Process and disturbance dynamics can vary greatly from process to process, and adjustments are available to compensate for these differences. A common misconception is that control loop tuning is the adjustment of controller tuning parameters and that the controller tuning parameters compensate for any and all process anomalies. What is often overlooked is the requirement to verify the appropriateness, the location, the installation, and the adjustment of *every* device in the control loop, including the equipment, field instrumentation, signal tranmission, controller, and final control element.

Controller tuning parameter adjustment, which is but one facet of the control loop tuning process, should usually occur after the other devices in the loop have been verified for appropriate operation. Even with the wide range of controller tuning parameter adjustment available, an inappropriate or misadjusted instrument in the control loop will cause the control loop to function in a less than optimal manner, despite the appearance of proper operation. In many loops, no amount of controller tuning parameter adjustment will adequately compensate for inappropriate equipment, instrumentation, or installation. Unfortunately, it is all too common to start control loop tuning by adjusting the controller tuning parameters without giving a thought to the remainder of the devices in the control loop.

It should be understood that many adjustments are available throughout the control loop and that these adjustments are present in the equipment, the sensor, the transmitter, the controller, and the final control element. These adjustments may not be described as such in an instrument instruction manual, especially in the case of field instrumentation, where location and installation effects can adversely affect performance of an instrument that, in turn, affects the operation of the loop.

Therefore, consideration must be given to *all* parts of the control loop, because any of these adjustments may be the factor that limits the operation of the entire control loop and, perhaps, the entire control system.

This chapter discusses the importance of the application, installation, and adjustment of field equipment as it relates to the control loop. Further insight into the various types of field equipment may be obtained from the references cited in the appropriate chapters.

## Control Loop Tuning Procedure

Control loop tuning should be performed in a logical procedure to ensure that all devices in the loop operate properly and that the loop performs satisfactorily in conjunction with the process. Failure to do so can result in less than optimal performance, given the investment in equipment. Note that the appropriate responses of the equipment and the instruments to manipulation are dependent upon the nature of the process.

PROCEDURE:

1. Prior to start-up, verify that the equipment and each instrument in the loop is appropriate for the application at hand and that the instruments intercommunicate through appropriate technology.

2. After the equipment is operating properly during start-up, verify the sensor, the transmitter, and the signal transmission performance at the input to the controller with the controller in manual mode. Similarly, check the operation of the final control element by manipulating the controller output with the controller in manual mode.

3. With the process operating, adjust the controller tuning parameters for automatic operation.

## Equipment Adjustments

Most mechanical equipment requires adjustment for proper operation. Attempts to start up instrumentation on equipment that is not functioning properly are usually futile. Be sure that equipment is in proper working order *before* attempting to make instrument adjustments.

## Sensor Adjustments

Few sensors have mechanical or electrical adjustments in the classical sense; however, their location and speed of response can be altered. Sensors should be located *WHERE THE PROCESS VARIABLE IS TO BE CONTROLLED.* This approach may seem trite, but many times the process variable is measured at an inappropriate location for any of a number of reasons.

### Example 6-1:

The temperature of the contents of a vessel should be measured in the liquid in the bottom of the vessel, that is, in the location most indicative of the liquid temperature. However, the thermowell and sensor may actually be installed in an existing well in the discharge piping to avoid the expense and inconvenience of welding a nozzle on the vessel. Therefore, the temperature is not being measured in the desired location. Aside from process considerations, such as the requirement that the liquid be flowing to continually transport liquid of the proper temperature to the point of measurement, additional complications may occur because of improper or inappropriate insulation.

**Dampening** In addition to appropriately locating the sensor, the sensor may require mechanical adjustments to make it more responsive to the process. When this is not possible, it may be desirable to employ another measurement technique.

### Example 6-2:

Temperature measurement of a pipe is achieved from a thermowell installed in a piping tee. Laboratory data confirms that the temperature measurement is in error. Examination of the installation reveals that the thermowell with a 3-inch extension is installed in a flanged tee located 12 inches away from the liquid flow. Therefore, the location of the measurement is in a stagnant pipe, the temperature of which is not indicative of the flowing liquid. The measurement will not be responsive to the process, not because it is performed in an inappropriate location in the process, but rather because the sensor is insulated from the process. This sluggish response can be remedied by installing a thermowell of appropriate length and, if additional speed of response is desired, applying a thermowell with an integral thermocouple.

At times, mechanical "dampening" may be necessary to make the sensor less sensitive to the process. For example, a block valve that connects the process to a pressure sensor may be partially closed to dampen the effect of line pressure pulsation caused by a pump.

## Transmitter Adjustments

The basic function of the transmitter in the regulatory control loop is to *MEASURE THE DESIRED PROCESS VARIABLE*, which, in the case of regulatory control, is the process variable that will be controlled. Due to the nuances of process measurement techniques and transmitter design, measuring the desired process variable is not as easy or as straightforward as it may seem.

***Transmitter Calibration Range***   Most transmitters allow adjustment (calibration) of the process zero and process span, which correspond to the zero and span of the transmitted signal, respectively. Assuming these adjustments are made to the correct specifications in a technically correct manner, they still may be *inappropriate with regard to their application to the process.*

The selection of an inappropriate calibration range typically occurs when the transmitter is calibrated for use under all possible operating conditions, including extreme conditions such as start-up and emergency shutdown. The use of a large measurement range can cause significant process variations that occur near normal operating conditions to result in small measurement signal changes. Therefore, the process can appear to be under control when the process variable is displayed using the transmitter range, such as in the case of a controller or recorder, potentially masking the importance of the variations.

### Example 6-3:

A product in solution can be concentrated by boiling the solution to remove some or all of the solvent. Temperature is often used to infer the product composition of the concentrated product and control heat input. See Figure 6-1.

The composition of a hypothetical product composition as a function of temperature is:

| Temperature (degrees Celsius) | Composition (percent) |
|---|---|
| 150.0 | 45 |
| 150.5 | 50 |
| 151.0 | 55 |

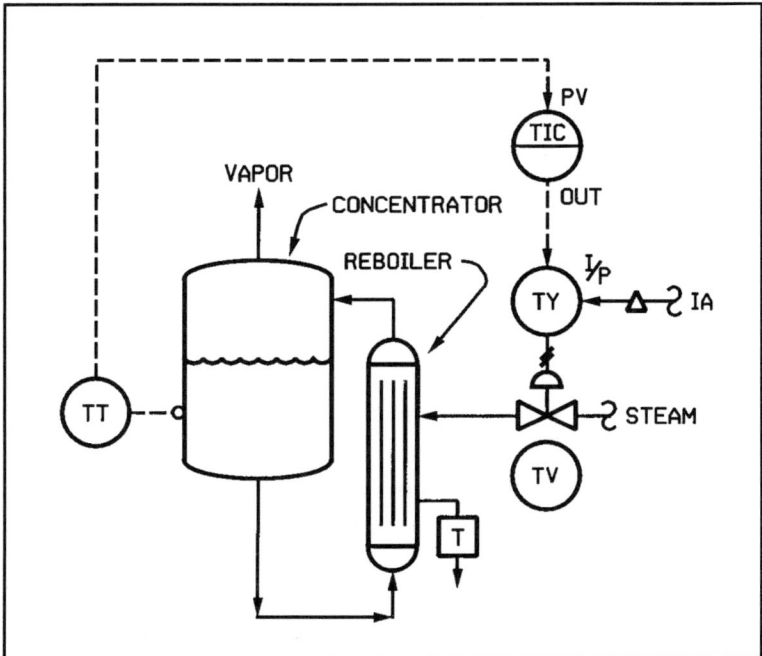

**FIG 6-1.** Concentrator temperature control

To handle start-up, operating, and over-temperature conditions, it would be reasonable to specify a temperature transmitter with a calibration of 0-200 degrees Celsius. Therefore, a one degree Celsius variation in temperature will affect product composition by 10 percent.

However, the one degree variation is only 0.5 percent of the transmitter range. Examination of this variation on a temperature recorder or indicator with a 0-200 degree Celsius range will appear to be a straight line, indicative of a process that is under control, despite the significant composition variation. Further, measurement accuracy typically deteriorates as the calibrated range increases, potentially causing process control problems during normal operation.

Application of a "smart" transmitter that is capable of accurately measuring the desired temperature range (perhaps 145 155 degrees Celsius) could be used to implement the regulatory control loop shown. Notwithstanding more sophisticated approaches, the "smart" transmitter approach, even if in addition to the 0-200 degree Celsius transmitter, would probably be cost effective due to its ability more accurately to infer composition and better correct for fluctuations in composition.

***Dampening***  Many transmitters have dampening adjustments to minimize the effect of process noise on the output signal. In most applications, the transmitter dampening adjustment is initially set to its default value. During start-up and operation, transmitter operation can be assessed and appropriate dampening adjustments can be made sparingly to maximize transmitter responsiveness while maintaining an acceptably low signal-noise level.

In addition, some flowmeters have a transmitter dropout feature that sets the flowmeter output to zero when the flow is below a given percent of full scale (typically 1-2 percent) to minimize noise and false signals at low flows.

***Signal Transmission***  For all practical purposes, electronic signal transmission is instantaneous. However, errors can occur when wires are improperly connected, the insulating property of the wire is compromised, when ground loops occur, or when signal strength is lost because of excessively long wiring. Digital communication is not prone to errors due to loss of signal strength per se, but a loss of communication will occur if signal strength is sufficiently low.

It should be noted that pneumatic transmission systems inherently exhibit dampening due to the finite time required for the transmitter (and repeaters) to supply compressed gas through the pneumatic tubing system. This effect may significantly dampen the response of the transmitter system and mask significant process fluctuations. In addition, tubing leaks can cause inaccurate measurements by shunting some of the signal (pressure) to atmosphere.

## Controller Input Adjustments

The transmitted measurement signal becomes the input to the control device, and the most prevalent input adjustment of a control device is input dampening. Depending on the technology, this adjustment may be made by adjusting a valve, a potentiometer, or a filter time constant. Whereas the transmitter dampening adjustment is used to reduce the signal noise level while maintaining transmitter responsiveness, the input dampening adjustment can be used to dampen process transients that are caused by actual process deviations that may affect the control. It is usually preferable to discover and remedy the cause of the transients than to dampen them out, and input dampening should be used sparingly.

In most applications, the input dampening adjustment should be initially set to its default value. During start-up and operation, control loop operation can be assessed and appropriate dampening adjustments can be made (if necessary) to stabilize the control loop.

Provided that sufficient adjustment is available, either the transmitter dampening adjustment or the input dampening adjustment can be made to satisfy both the measurement and control requirements. Note, however, that such an adjustment (especially if made at the transmitter) may mask important transients that could be caused by instability elsewhere in the process (and subsequently corrected). Individual adjustments that utilize the objectives described above will (at least) provide a signal that reflects the actual process variable, with noise filtration performed at the transmitter (probably utilizing a superior algorithm developed by the transmitter/sensor manufacturer), which allows transients to be measured by the control system and analyzed.

### *Example 6-4:*

A pressure transmitter is used to measure the pressure of a gas in a vessel whose pressure is controlled. See Figure 6-2.

Due to the pulsation caused by the gas compressor, the undampened measurement and recording are inherently noisy, as reflected by a pressure recording that is a relatively wide band that varies with time. The transmitter dampening can be

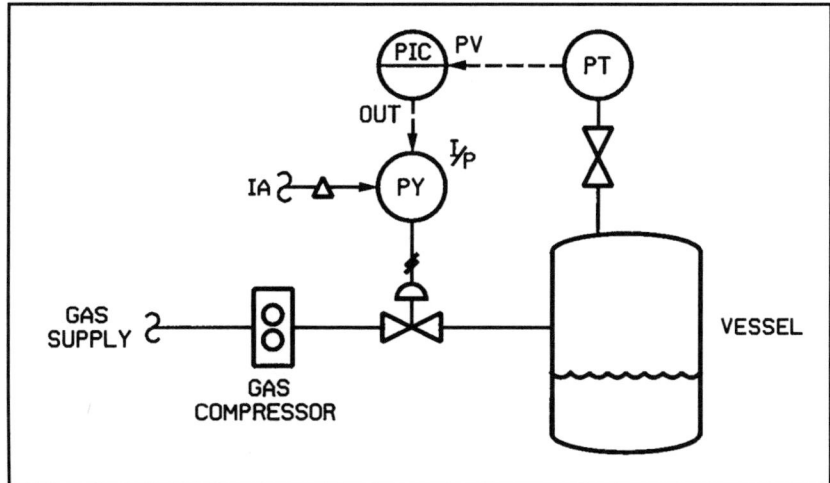

**FIG 6-2.**   Gas pressure control

increased to dampen the measurement (only), thereby reducing the pulsation noise and the width of the recorded band. The varying pressure that remains may be caused by process considerations or, perhaps, by valve hysteresis.

The input dampening adjustment is still available if required, but it is preferable to correct the cause of the variation before increasing the input dampening, which can mask the variation.

It should be noted that had the transmitter or input been excessively dampened, the actual pressure variations may not have been apparent, and the effects of instability elsewhere may have been masked. Further, a pneumatic, signal-transmission system, with its inherent dampening effect, may excessively dampen the measurement system, that will create the same result.

## Final Control Element Adjustments

In most industrial processes, the final control element is a control valve that is used to manipulate flow. Due to reduced costs, energy savings, and other considerations, variable speed drives are being applied to industrial processes to manipulate flow by varying the amount of mechanical energy put into the system.

***Control Valves*** In many industrial applications, the pneumatic spring return actuator provides the mechanical energy to operate the control valve. It should be clearly understood that the pneumatic signal that is applied to this actuator is nothing more than that—a pneumatic signal applied to the actuator. The value of this signal is often misunderstood to be equal to the position of the control valve. Factors in addition to the signal will affect valve position, such as I/P converter accuracy, process conditions (including differential pressure), control valve hysteresis, and valve adjustments.

The addition of a positioner (which can be considered a valve adjustment) introduces a local, valve-position control loop with the valve signal as its set point. This loop manipulates the pressure to the actuator in such a manner that the valve is moved to the valve position that is represented by the signal. Utilizing a positioner, the signal is indicative of valve position, because the effects of the process conditions and valve hysteresis are reduced.

Electrically operated control valves are used extensively in industry and exhibit similar characteristics in that they may or may not contain positioning circuitry. It should be noted that without positioning circuitry, the valve position may change as a result of process conditions.

Mechanical adjustments can be made to control valves, including adjustment of the packing, the actuator stroke, the supply pressure, and the like.

### *Example 6-5:*

The liquid flow control loop feeding a tank is operated in the manual mode. (See Figure 6-3). In this case, the manually adjusted controller output sends to the valve a signal that is representative of the air signal to the control valve actuator diaphragm.

This flow is known to be difficult to manipulate, because varying the signal from 30 to 75 percent does not affect the flow. Examination of the valve reveals that the valve packing was tightened excessively after an apparent packing leak, resulting in excessive control valve hysteresis.

If the problem persists after making the proper packing adjustment (after repacking, if required), a positioner can be installed to reduce the effects of hysteresis. If the packing must be tightened excessively to reduce leakage, the installation of a control valve that utilizes a different design may be considered.

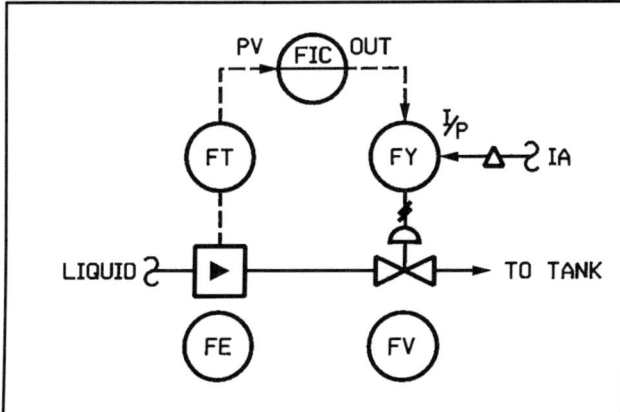

**FIG 6-3.**   Liquid flow control

***Variable-Speed Drives***  A number of variable-speed drive adjustments are available to the user that may be manipulated with potentiometers or digitally, depending upon the design. Parameters commonly considered for adjustment for control purposes include the minimum and maximum speeds, ramp-up and ramp-down times, and skip frequencies. These adjustments set the range of operating speeds, the rates at which motor speed can be increased and decreased, and frequencies at which the drive will not be allowed to operate. Variable-speed drives exhibit virtually no hysteresis similar to that caused by control valve packing, because the motor makes small speed adjustments electrically.

Minimum and maximum speeds of liquid positive-displacement equipment can usually be set at 0 and 100 percent, respectively; however, gas positive-displacement equipment may be required to operate above some (non-zero) motor speed to protect the equipment from internal damage. The minimum speed of centrifugal equipment that must overcome a static head should be set to a speed at which the static head generated at zero-flow conditions is somewhat less than the static head that must be overcome.

The ramp-up and ramp-down times determine the responsiveness of the equipment to a signal change. These settings usually can be left as they come from the factory, but some equipment requires more gentle speed variation in order to protect the equipment and/or not to overload the drive. Increasing the responsiveness of the motor should be performed during start-up or during operation when process conditions warrant and excessive equipment wear will not occur.

In addition, certain types of equipment require the setting of skip frequencies that inhibit the drive from operating at those frequencies that can cause damage due to resonance.

## Summary

In order for a control loop to function to its fullest, all components of the control loop must be properly adjusted, including the field equipment. Failure of the field transmitter to faithfully measure and respond to process conditions compromises the effectiveness of the control loop.

## ❖ Chapter 7

# *Regulatory Controller Features*

The regulatory controller compares the measurement with the desired set point that has been established by the operator; its output manipulates the final control element—typically a valve. Before adjusting the controller tuning parameters, a number of controller features must be enabled and adjusted to ensure that the controller will function properly. Controller tuning performed before setting up the controller may be a frustrating and futile effort.

## Anti-Reset Windup

In many processes, upset conditions can cause the controller output to saturate at one of its extremes. The process is "out of control" in the sense that the process variable cannot be returned to its set point by manipulation of the final control element within the limits of the controller output. The attempt to bring the process to its set point may require that the controller output remain at an extreme for an extended period of time, causing it to "wind up", that is, to saturate to a value in excess of its nominal extreme.

### *Example 7-1:*

The average steady-state flows in a condensate system are shown in Figure 7-1. These flows may be typical of the operation but are not representative of upset conditions.

Consider the sequence of events initiated when the de-aerator makeup pump turns off. A few minutes later, the de-aerator level alarm (not shown) warns the operator of a lower than normal water level in the de-aerator. The operator determines the cause of the problem and energizes the backup de-aerator makeup pump.

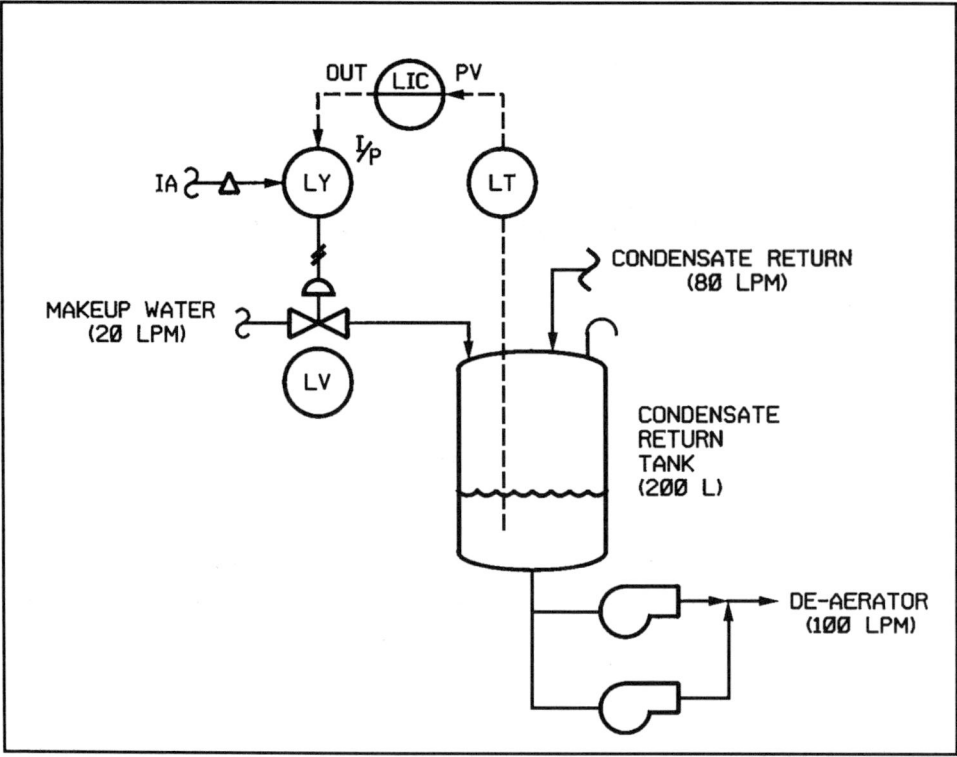

**FIG. 7-1**   Condensate return system

Assuming that the tank was half full prior to a 5-minute loss of pumping, 400 liters of condensate returned to a tank with only 100 liters of usable volume. Due to the rise in level, the makeup water valve closed shortly after the pump turned off, in an unsuccessful attempt to lower the liquid level in the tank. By the time the backup pump was energized, more than 300 liters of condensate had overflowed the tank.

During this sequence of events, the rise in tank level caused the controller output to go to its minimum value in order to close the valve. Although the minimum signal value is normally considered to be 4.00 mA, the integrating action (reset) of the controller may cause the actual controller output to be as low as 2.5 mA, the value at which the controller output electronics saturate.

When the tank level falls after the pump turns on, valve response is delayed until the controller output integrates from 2.5 mA to above 4.00 mA, causing a delayed, makeup water response that can result in a more pronounced drop in tank level and subsequent loss of pump suction.

A controller with an anti-reset windup feature would limit the controller output to between 4.00 mA and 20.00 mA, which would allow the valve to respond without waiting for the controller to come out of saturation. Similarly, controllers can exhibit reset windup at full scale with as much as 27 mA.

While this example shows the effects of reset windup, it also illustrates the potential need for a larger condensate tank. However, more sophisticated controls and alarms may allow use of the existing tank or complete elimination of the tank. The process control engineer can aid in making this determination.

---

Reset windup is an inherent feature of many controllers, but it is most prevalent in pneumatic and analog electronic controllers in which special anti-reset windup circuits must be specified to limit the controller output signal to its nominal values. Digital controllers and distributed control systems do not require special circuitry and can perform this function in software.

# Bumpless Transfer (Set Point Tracking)

The controller output that manipulates the final control element can be adjusted manually by the operator or automatically by the controller, depending upon whether the operator has selected the manual or the automatic mode of the controller. During switching between these modes, conditions can arise that cause improper set points and outputs to be generated.

When the controller is in manual, the operator can directly manipulate the output. At this time, the set point can be changed, but these changes will not affect the output. Although the loop may be stabilized in manual, when the operator switches the controller to automatic, the controller algorithm adjusts the output in response to the difference between the set point and the process variable. This can cause an large upset to the process, if the operator fails to set the set point equal to the process variable *before* switching to automatic.

In some applications it is advantageous to unburden the operator from the inconvenience of having to set the set point equal to the process variable, which can be accomplished through the use of a "bumpless" transfer feature. During the switch from manual to automatic, this feature causes the set point to equal the process variable before the control algorithm manipulates the output. As a result, the control loop is switched from manual to automatic without a bump in the output signal. In some controllers, the set point may follow the process variable in manual (track the set point), effectively yielding the same result. It should be noted that while the bump is avoided, the operator must adjust the set point to the desired value. This, however, is a function with which the operator should be familiar.

### *Example 7-2:*

Consider the start-up of a drying oven in which changing the oven temperature rapidly can cause the refractory to fail. It is common for the operator to incrementally raise the firing rate over time by manipulating the fuel valve with the controller in manual. When the temperature is about half of the desired operating temperature, the operator can switch the controller to automatic and let the controller do the work.

This may seem straightforward, but the controller set point may have been left at the normal operating temperature, and the operator may have failed to set the set point equal to the process variable. In such a case, when the controller is switched to automatic the set point may be hundreds of degrees higher than the oven temperature, which will cause the fuel valve to open fully, thus abruptly raising the oven temperature, and potentially causing refractory failure.

The addition of the bumpless transfer feature would unburden the operator and minimize the risk of refractory damage.

It should be noted that, whereas the bumpless transfer feature would seem advantageous, it is not applicable to all processes. For example, the start-up procedure of a flow control loop may be to switch the controller to automatic with the proper set point, allowing the controller to manipulate a valve to achieve the desired flow. Application of bumpless transfer to this process would burden the operator with the additional work of setting the set point to its proper value after switching to automatic or manually stabilizing the loop near the proper set point before switching to automatic control.

## Control Dead Band

When the process variable is near the set point, it is possible for measurement noise to manipulate the controller output and the final control element, which will cause the process variable to oscillate near its set point. Some controllers allow a control dead band to be set around the set point; control action stops when the process variable is within this dead band, and the controller algorithm maintains the controller output at its last value.

This feature can be very helpful in minimizing the effects of measurement noise on the process, because the noise is ignored when the process variable is within the dead band. Note that the measurement signal need not be overdamped (as would be done to reduce oscillations, if this feature were not available), and the controller can be responsive to relatively rapid process changes outside the controller dead band. To minimize the effects of noise, the controller dead band should set as small as practical.

The control dead band feature can also be used to better control processes that do not require tight control. This may seem contradictory, but consider the control of the condensate tank in Figure 7-1, which serves to absorb batches of condensate returned after it is accumulated in local tanks. The condensate tank level rises abruptly when a batch is returned and falls gradually as the condensate is used while there is no return.

Tight control would require that the level control valve be closed (open) when the level is above (below) the setpoint. Opening the control valve fully when the level is below the set point (between batches) may put an operational strain on the makeup water supply.

In this process, there is little need to tightly control level because the condensate needs of the de-aerator are satisfied as long as condensate is available; that is, what matters is not tank level but rather that condensate is available and does not overflow the tank. Therefore, the condensate tank can be allowed to fluctuate between 40 percent level to ensure sufficient pump net positive suction head (NPSH) and 70 percent level to allow volume for returning batches of condensate.

Note that it is not sufficient for condensate to be present in the tank. Because the condensate may be near its boiling point, a relatively low tank and high pump NPSH requirement dictates maintaining the higher condensate tank level to avoid cavitation and subsequent pump damage. This additional requirement reflects the process requirement that the control of condensate tank level is not the process, but rather is the means by which the process is managed to allow the condensate be delivered to the de-aerator.

Recognizing that tight control is not required, a set point of 55 percent and a controller dead band of 10 percent can be utilized to allow the level to fluctuate with the process while keeping the level under control, that is, between 40 and 70 percent level. When in automatic, the controller output will maintain its last output when the tank level is between 45 and 65 percent. Outside of these limits, the controller will manipulate the control valve until the level returns within the limits, again holding its last output.

Eventually, these excursions will cause the controller to reach an output at which the level does not exceed the controller dead band, sending a constant controller output to the makeup water valve even though the batches are returning and the level is fluctuating accordingly. With the exception of an upset condition, the makeup water flow will be relatively constant and will not overload the makeup water system.

## Controller Range

The measurement range of the process variable determines the range over which the controller can control. Whereas previous generations of controllers required that the measurement range and the controller indicator be identical, controllers that utilize digital electronics allow the controller faceplate to indicate a range different from that of the measurement, as well as a digital indication of the process variable. The adjustment of a narrow indication range in which the process normally operates allows more accurate set point adjustment and easier detection of actual process deviations.

### Example 7-3:

A temperature controller is used to control furnace temperature at 1000 degrees Celsius. See Figure 7-2. The temperature transmitter range (and hence the controller range) is 0-1500 degrees Celsius to measure start-up and upset conditions in addition to the normal operating temperature of approximately 1000 degrees Celsius.

If the controller indicated 0-1500 degrees Celsius, the loop would appear to be under control even if process fluctuations of 50-100 degrees Celsius were present. However, if the controller were adjusted to indicate from 900-1100 degrees

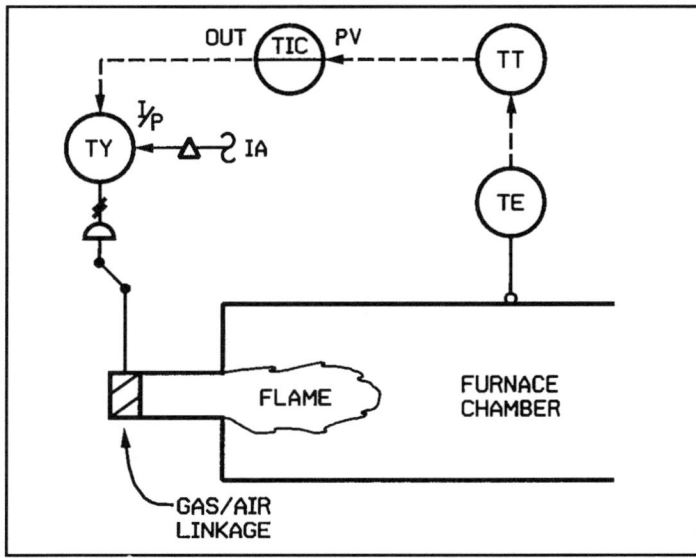

**FIG. 7-2**   Simplified furnace chamber temperature control loop

Celsius, process fluctuations of 50-100 degrees Celsius could easily be detected. Further, the digital indicator on the controller faceplate would provide the operator with digital temperature information at all times, including start-up and upset conditions.

# Direct/Reverse Control Action

In automatic operation, the controller output either increases to increase the process variable (direct action) or decreases to increase the process variable (reverse action).

### *Example 7-7:*

Figure 7-3 is a simplified drawing of typical flow controls (only) for a distillation column reflux tank. Note that the reflux flow valve fails open, and the take-off flow valve fails closed. This combination implements the process requirement to reflux all material in event of an air loss or instrument signal loss.

The reflux flow controller is reverse-acting, that is, the controller output must decrease (opening the fail-open flow valve) to increase the process variable. The product take-off flow controller is direct-acting, that is, the controller output must

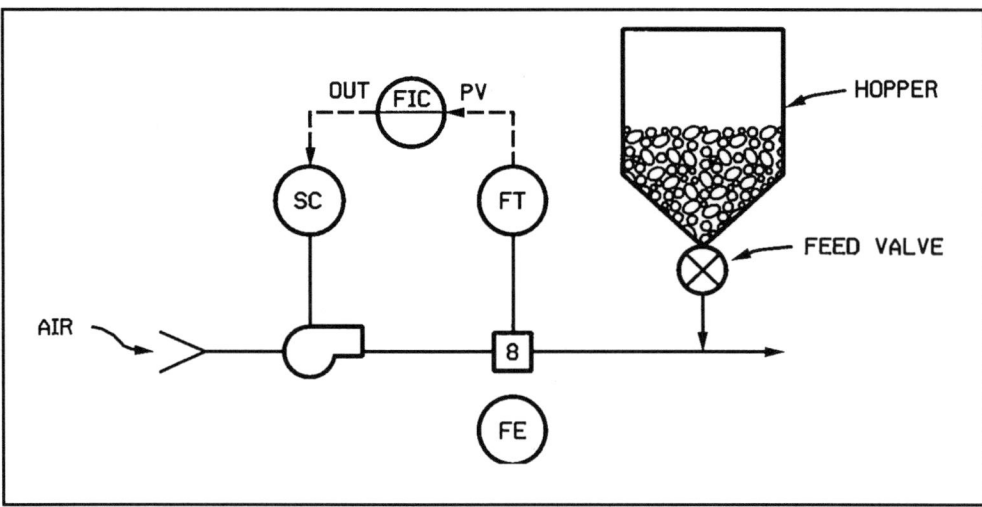

**FIG. 7-3**    Air flow control.

increase (opening the fail-closed flow valve) to increase the
process variable.

## Failure Modes

Despite the many potential controller failure modes, only
a small number of practical failure modes are considered by
the user. The most likely is a controller power failure that affects the control loop when power is removed and again when
it is restored.

***Loss of Power***    The analog output of a controller has a polarity, in the sense that the minimum controller output signal
may cause the final control element to be at its minimum or
maximum, depending upon the application and the nature of
the final control element. When power to the controller is lost,
safety considerations are usually the overriding concerns used
to determine the failure position of the final control element
and, in turn, the polarity of the controller output signal. On
the other hand, safety and economics may be the driving
forces in determining how the controller is to respond after
power is restored.

Whenever possible, the final control element should move to
its failure position when the controller output signal is at its
minimum. Further, "fail-safe" operation requires that a failure
of the controller signal or the final control element power
source will cause the final control element to move to its fail-

ure position without the use of external power sources, such as electricity, instrument air, and the like. When "fail-safe" operation is desired, it is important to verify that the final control element moves to its failure position upon the loss of *any* external power source.

Even though "fail-safe" operation may not be required in many applications, the integration of this concept into the equipment adds operational security and creates less confusion once it is understood. In addition, it does not increase cost.

In a "fail-safe" installation, a loss of power to the controller results in a loss of the controller output signal, which will cause the final control element to move to its failure position regardless of whether other power sources have failed. In addition, failure of the external power sources that feed the final control element will cause the final control element to move to its failure position independent of the controller output signal.

Most electrically operated valves cannot be operated as "fail-safe" because a loss of power or loss of signal will cause the valve to remain in its last position, although some actuators will move the valve to a designated failure position on loss of signal. Similarly, variable speed drives will turn off if power is lost, but many may be configured to appropriately respond to a loss of signal. Electrically operated valves are usually not applied in processes that require fail open (F.O.) or fail closed (F.C.) "fail-safe" operation in order to maintain safety. Variable speed drives inherently stop when power is lost and can be utilized in "fail-safe" applications that require the final control element (pump) to fail off.

***Restoration of Power***  When power is restored to the controller, the final control element should move to the position that corresponds to the controller's present output signal. However, all of the external sources to the final control element may not be available, and the final control element may remain in its failure position. In either case, the controller is typically in manual, but may be in automatic if the technology and the design allow.

In most processes, it would seem to be acceptable for the controller to be in manual, generating a zero output signal after power is restored. The operator could then manually adjust the final control element or raise the set point in automatic. This presents a potential problem in some applications, because the operator may be faced with the simultaneous start-up of many control loops, in addition to the major process upset that occurred during the momentary controller power loss. Therefore, after a power failure, it may be easier, safer, and more economical to allow as many control loops as practical to startup with the controller in automatic with a preset or the last set point, or in manual with a preset controller output or the output prior to losing power.

A number of considerations must be weighed in the decision regarding the controller output and automatic/manual status after restoration of power. These considerations are process-dependent and should be investigated in a formal hazard review. They include such items as: (1) the consequences of the final control element overshooting to its maximum; (2) or not moving immediately after restoration; (3) or not being in automatic, and the like. After the restoration of power, the decision to configure the instrument to force the controller to manual or to return the controller to its mode of operation prior to power failure depends upon knowledge of the process, and how the measurement and final control element are integrated into that process.

***Other Failure Modes***  It should be noted that other likely failure modes may be found by evaluating the controller design and technology. Examples include the effects of pneumatic tubing leaks, dirty instrument air, electrical noise, etc.

## Event-Related Controller Features

Controllers have traditionally operated in the analog domain, but controller hardware and software developments have enabled internal logic functions to influence the controller in more sophisticated ways than those traditionally achieved using external logic circuits and relays. The event-driven controller responses described are typical of many digital regulatory controllers, but some digital regulatory controllers use configuration and programming techniques to allow customization of these features.

***Enabling Switches***  Controllers may contain a number of hardware or software switches that enable and disable numerous controller functions such as alarm types, bumpless transfer (set point tracking), hard output limits, output tracking, remote set point control, and the like. It is not only important to enable the desired functions but also to verify that they are disabled to ensure that the remaining functions do not interfere with the desired features.

### Example 7-4:

The default of a controller database may enable the bumpless transfer function. Other desired functions may be enabled for a specific application, however, if bumpless transfer is not desirable in this application, this switch must be disabled to avoid operational problems.

---

***Manual/Automatic Switch*** In some applications, it is desirable for the controller to switch from automatic to manual and vice versa as the result of a specific event. This feature could be used to effectively locate the automatic/manual switch remote from the controller so that the operator could switch the controller while directly observing the process. In other applications, specific process conditions could trigger the switch.

Care should be taken to determine whether the condition or a change in condition causes the controller to switch. This aspect of the manual/automatic switch should be investigated while it is appropriately installed to match the requirements of the process, especially with regard to the ability of the operator to manually manipulate the final control element after the switch to automatic—an ability that may or may not be desirable in a given application.

***Output Preset Value*** Some applications require that the controller be forced from automatic to manual as well as to a preset output when a specific condition is present. When the condition is not present, the controller switches from manual to automatic and, after the switch, usually (but not always) allows the operator full control of the loop, including the ability to switch the controller to manual.

### Example 7-5:

An air flow control is used to pneumatically transport a powder (see Figure 7-4).

When the feed valve is off, the flow controller is forced to manual, and the fan is forced to a preset speed in order to conserve energy. When the feed valve is operating, the controller is switched to automatic in order to control the air flow at its set point. Note that bumpless transfer is not desirable in this application.

Controller responses to the condition and to a change of condition must be investigated to ensure that the responses are compatible with the process (preferably minimizing the number of actions required by the operator).

**Output Tracking**  In some processes, it is desirable for the controller to be forced to manual and its output forced to the value of one of its analog inputs that represents a signal from another source.

### Example 7-6:

This feature can be applied to allow the control system to bypass a controller and directly manipulate its output when a constraint is reached (see Figure 7-5).

In this application, a gas is fed to the reactor at a flow rate set by the operator. If the flow rate is excessive, there may not

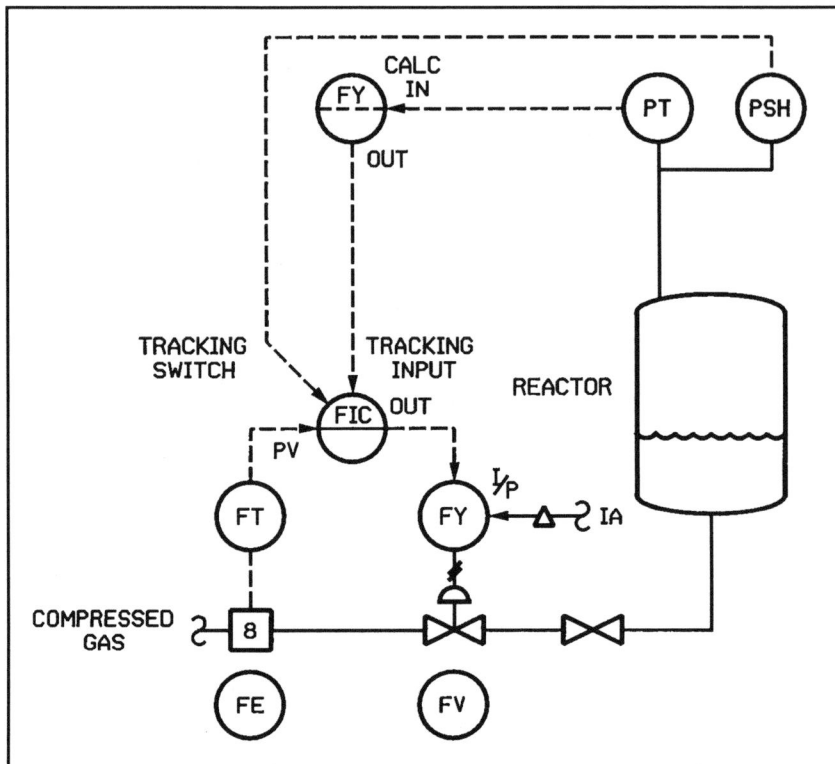

**FIG. 7-4**   Reactor gas flow control with output tracking.

be sufficient time for the gas to react with the liquid in the reactor, and the reactor pressure can increase because of an accumulation of unreacted gas. Should this occur, the reactor pressure switch will sense high pressure, send a contact closure to the flow controller that will cause the flow controller to cease automatic operation, and output the calculated value at its tracking input that is a function of the reactor pressure. When operating in this manner, the controller is effectively bypassed and its output is manipulated directly by the tracking input.

## Output Limits

Output limits can be used when it is desirable to limit the controller output in order to avoid unsafe, unstable, or unde-

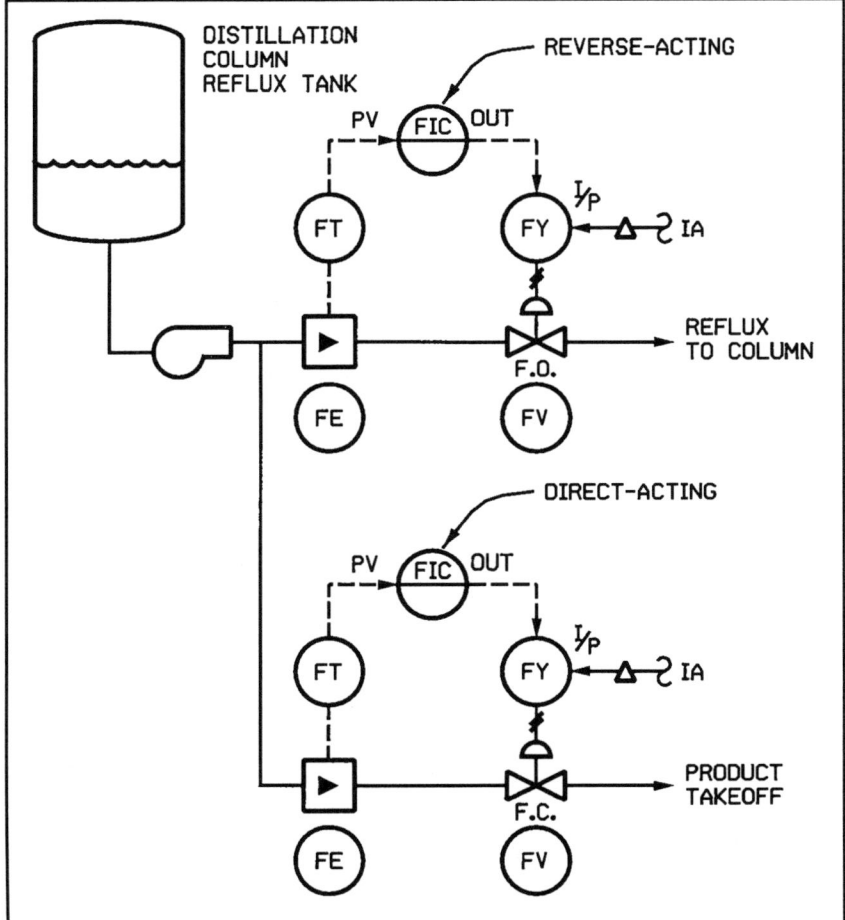

**FIG. 7-5**   Simplified reflux and product takeoff flow controls (only).

sirable operating conditions. It should be noted that some controllers limit the output while in all modes of operation; others limit the controller output only when the controller is in automatic operation. A switch may be available in the controller to enable or modify these functions.

### *Example 7-8:*

The natural gas flow through the natural gas flow valve of a multifuel boiler cannot sustain combustion during operation under low-load conditions because the valve closes to less than 25 percent open. To avoid flaming out the boiler under these conditions, a natural gas controller low-output limit can be adjusted to approximately 28 percent. This output limit should be made active only during automatic operation to allow the complete removal of natural gas when the controller is in manual and the burning of other fuels in sufficient quantity to sustain combustion.

## Remote Set Point

In many applications, the controller set point is adjusted by the operator. This need not be the case, however, because some controllers allow external signals to adjust the controller set point and, in addition, allow the external signal to be scaled, using ratio and bias adjustments. Enabling the remote set point feature also enables a remote/local set point switch on the controller faceplate that allows the operator to switch between the local (operator-adjusted) set point and the remote set point.

### *Example 7-9:*

Figure 7-6 illustrates a flow control loop whose set point is generated by a signal from the computer when the controller is in remote set point operation. Note that the operator can put the controller in local control and adjust the set point, or put the controller in manual control to directly manipulate the controller output.

## Set Point Limits

Set point limits can be used to minimize the ability of an operator (or control system) to excessively raise or lower the

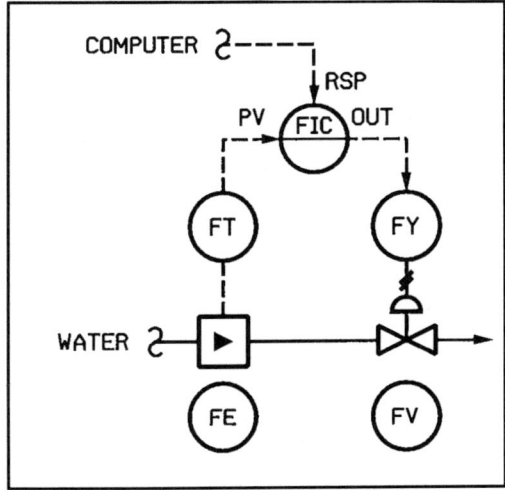

**FIG. 7-6**    Flow loop with remote set point

controller set point. While the set point limits may be active at all times, the limits will have no effect on the process when the controller is operated in manual because the operator directly manipulates the controller output (independent of the set point).

### *Example 7-10:*

The natural gas flow through the natural gas flow valve of a multifuel boiler can exceed the maximum firing rate of the boiler when the natural gas flow is more than 75 percent of full scale. To limit the natural gas flow in automatic operation and minimize the possibility of overfiring the boiler, the natural gas controller high set point limit can be adjusted to approximately 75 percent of full-scale flow. Note that the operator can further increase the natural gas flow while the controller is in manual operation.

## Slew Rates

In some applications, abrupt manipulation of the set point or output can upset the process, especially in processes that react slowly to manipulation. To minimize process upsets, slew rate adjustments may be available to limit the speed with which the set point to the control algorithm can be changed (regardless of the abruptness of operator actions) and the speed with which the controller output is changed by the control algorithm. The output slew rate may or may not be active when the controller is operating in manual.

### Example 7-11:

During start-up, it was noted that when the set point of the level control loop shown in Figure 7-1 was adjusted abruptly by 10 percent, the level would oscillate for a period of time before reaching a stable steady state. It was further noted that adjusting the level in 1 percent increments produced negligible oscillation. A set point slew rate adjustment could be made to this controller to allow the control algorithm to gradually be affected by the operator's abrupt 10 percent set point change.

## Totalizer

Some controllers have integral flow totalization capability. Controller adjustments must be made to properly scale the totalizer to the flow signal and allow for its display. This feature may be useful in some applications, but the inability of the controller to retain the total during a power failure may preclude its use in some controllers.

## Summary

Regulatory controllers should be configured to minimize hardware costs and increase safety and reliability. These features should be thoroughly understood during the control loop design and implementation phases of a project.

## ❖ Chapter 8

# *PID Control*

The regulatory controller is a device with a purpose of maintaining its input [process variable] at a desired value by manipulating a final control element, typically a control valve. The controller output that manipulates the final control element to achieve this end is generated by a mathematical algorithm that compares the process variable with the set point. The most common controller algorithm used to determine the controller output is the PID algorithm.

PID regulatory controllers contain hardware or software that implements proportional, integral (reset), and derivative (rate) control. The controllers described herein contain all three control algorithms, but individual controllers may delete some algorithms (not all are necessary for all control loops, and some may even be detrimental to the operation of certain control loops). Equipment costs of hardware implementations can be reduced by reducing the number of algorithms. Faster algorithm run time is achieved when fewer software algorithms are implemented.

The PID controller generates an error signal by taking the algebraic difference between the process variable and the controller set point. This error signal, which may be positive or negative, is used to implement the PID control algorithm.

Note throughout the following discussion that if the error is zero (that is, the process variable and set point are equal), no control action is taken, and the controller output will maintain its last value.

## Proportional Control

Proportional control action modifies the controller output by an amount that is proportional to the error signal (see Fig. 8-1). On the left side of Figure 8-1, before the step change, the process variable and the set point are shown to be equal in a

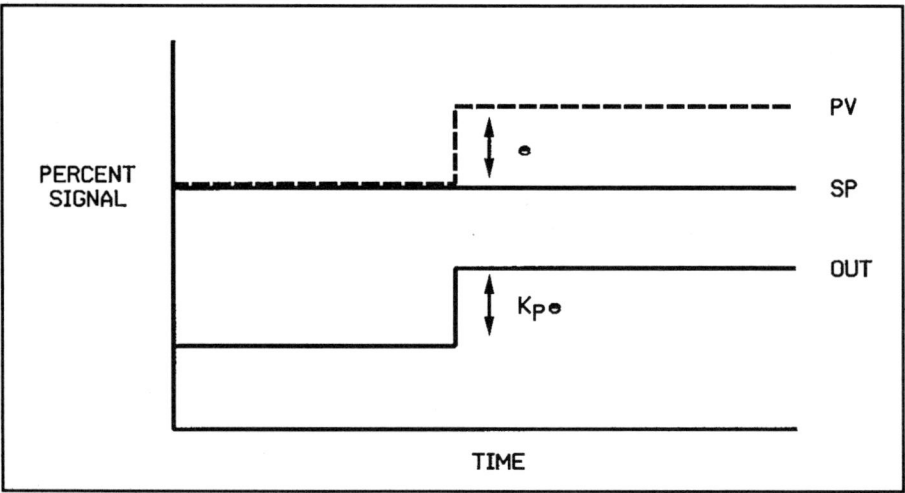

**FIG. 8-1**    Response to step process variable change—proportional control

controller with proportional-only control. As such, the output remains at its previous value because no control action is necessary.

When the process variable changes, an error (e) develops, at which time the proportional controller algorithm modifies the controller output:

$$OUT(t + 1) = OUT(t) + (K_p \times e) \tag{8-1}$$

where $K_p$ is the proportional controller gain and t is time. The illustrated controller response is a step output change. If the controller gain ($K_p$) were unity, the process variable and the output step changes would be identical. Because unity controller gain is not appropriate for all loops, the controller gain ($K_p$) must be adjusted to match the process.

Proportional band (PB) is often used to describe the proportional controller adjustment, where

$$PB\ (\%) = 100/K_p \tag{8-2}$$

The numeric value of proportional band is equal to the error as a percentage of the process variable range that will result in a 100 percent output change.

### *Example 8-1:*

A proportional controller with a controller gain ($K_p$) of 2 would have a proportional band (PB) of 50 percent. An error signal of 50 percent of the process variable range would cause a 100 percent change in controller output.

Note that the above discussion is intended to illustrate the concept of proportional control by assuming a step process variable change that results in a constant error signal and a constant controller output. In practice, the proportional control algorithm will be adjusting the controller output in proportion to the instantanious error signal which will not be constant in the great majority of applications.

## Integral Control

Integral control action modifies the controller output by an amount that is proportional to the integral of the error signal. Integration is a function of calculus that can be compared to taking the area under a curve (see Figure 8-2).

In another use of this concept, an integrator is used to "totalize" flow, which is nothing more than taking the area under the flow measurement curve (see Figure 8-2).

In the left part of Figure 8-3, before the step change, the process variable and the set point are shown to be equal in a controller with integral-only control. As such, the output remains at its previous value because no control action is necessary.

When the process variable changes, an error (e) develops, at which time the integral controller algorithm modifies the controller output:

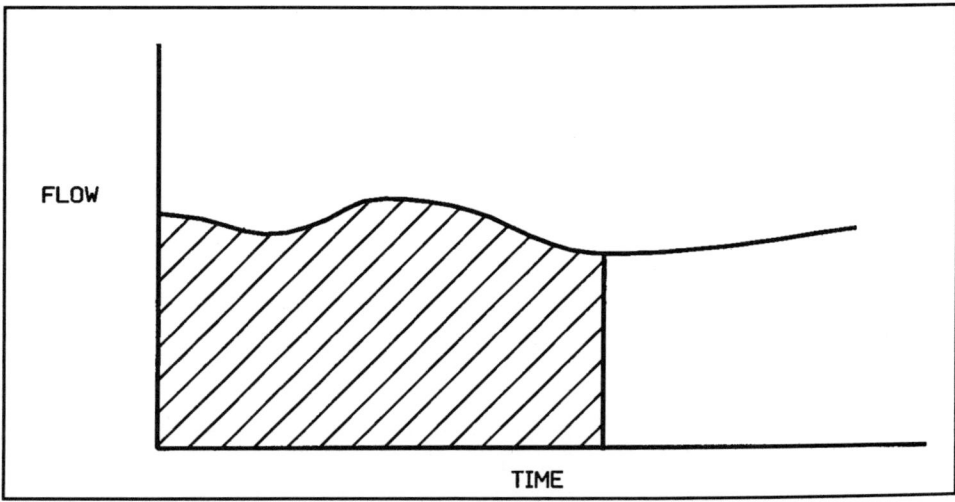

**FIG. 8-2**   Flow totalization performed by integration

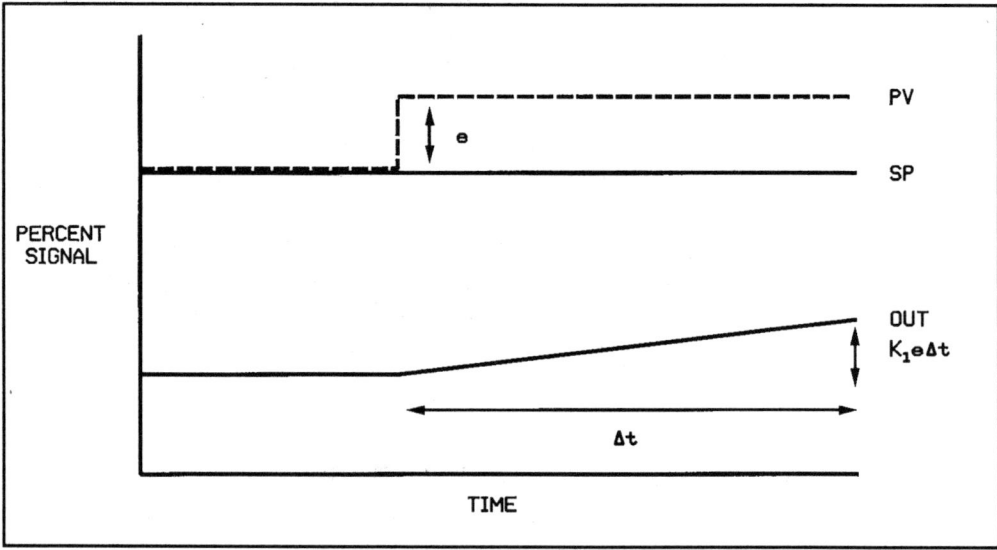

**FIG. 8-3**    Response to step process variable change—integral control (reset)

$$OUT(t + 1) = OUT(t) + K_i * \Sigma (e * \Delta t) \qquad (8\text{-}3)$$

where $K_i$ is the integral controller constant. The illustrated
controller response is a ramp output change that is proportional to the integral of the error, that is, the area between the
process variable and set point curves. If the integral controller
constant $(K_i)$ were unity, after one unit of time the output
value will have changed from its original value by an amount
equal to the error (e). Because a unity integral controller constant $(K_i)$ is not appropriate for all loops, the integral controller
constant $(K_i)$ must be adjusted to match the process.
The integral controller constant $(K_i)$, often referred to as the reset adjustment, is typically expressed in terms of minutes per
repeat, that is, the number of minutes necessary for the output
to ramp an amount that is equal to the error signal when the
error signal is held constant. Note that this adjustment may be
expressed as its inverse—repeats per minute.

### *Example 8-2:*

A integral controller with an integral controller constant $(K_i)$ of
2 minutes per repeat with a constant 10 percent error signal
would ramp its output at a rate of 5 percent per minute, reaching an output signal that is 10 percent higher than the initial
output signal in 2 minutes.

Note that the above discussion is intended to illustrate the concept of integral control by assuming a step process variable change that results in a constant error signal and a ramping controller output. In practice, the integral control algorithm will be adjusting the controller output by integrating the instantaneous error signal, which will not be constant in the great majority of applications.

## Derivative Control

Derivative control action modifies the controller output by an amount that is proportional to the derivative of the error signal. The derivative, which is a function of calculus used to determine the slope (or rate of change) of a curve, can be loosely defined as the change in value per unit time.

It should be noted that derivative control is used sparingly because measurement noise, small upsets, and set point changes can affect the error signal and potentially have a large effect on the controller output, inducing possible upsets and instability.

In the left part of Figure 8-4, before the step change, the process variable and the set point are shown to be equal in a controller with derivative-only control. As such, the output remains at its previous value because no control action is necessary.

When the process variable changes, an error (e) develops, at which time the derivative controller algorithm modifies the controller output:

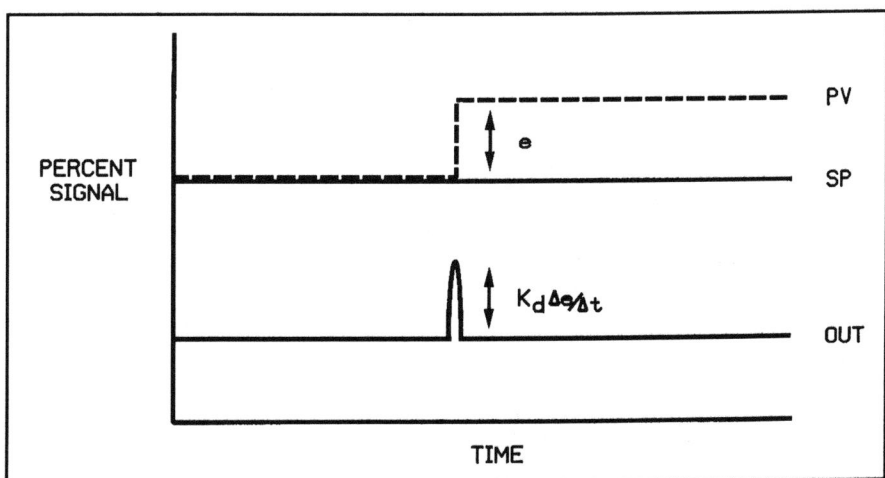

**FIG. 8-4** Response to step process variable change—derivative control (rate)

$$OUT(t + 1) = OUT(t) + K_d * \Delta e / \Delta t \qquad\qquad (8\text{-}4)$$

where $K_d$ is the derivative controller constant. The illustrated controller response is a pulse that is proportional to the magnitude of the derivative of the error, that is, the slope of the error curve. If the derivative controller constant ($K_d$) were unity and the slope of the error curve were constant, after one unit of time the output value will have changed from its original value by an amount that is equal to the rate of change of the error signal. Because a unity derivative controller constant ($K_d$) is not appropriate for all loops, the derivative controller constant ($K_d$) must be adjusted to match the process.

The derivative controller constant ($K_d$), often referred to as the rate adjustment (referring to an algorithm that utilizes the rate of change of the input), can be expressed in terms of minutes, that is, the number of minutes necessary for the output to change an amount equal to the rate of change of the error signal when the rate of change of the error signal is held constant.

### Example 8-3:

The process variable of a derivative controller is ramping at a constant rate of 1 percent per minute. With a derivative controller constant ($K_d$) of 10 minutes per repeat, the controller output will change 1 percent in 10 minutes, or 0.1 percent per minute.

Note that the above example is intended to illustrate the concept of derivative control by assuming that the error is changing at a constant rate. In practice, the derivative control algorithm will be adjusting the controller output by calculating the rate of change of the instantaneous error signal, which will not be constant in the great majority of applications.

***Derivative on Process Variable*** In many applications, the derivative control algorithm using the error signal described above can adversely affect the controller output when the set point is adjusted, unless the operator puts the controller in manual when the change is made or changes the set point slowly. Some controllers have the option of ignoring error changes that are generated by set point changes improving the functionality of the controller and minimizing the probability of the operator causing an upset.

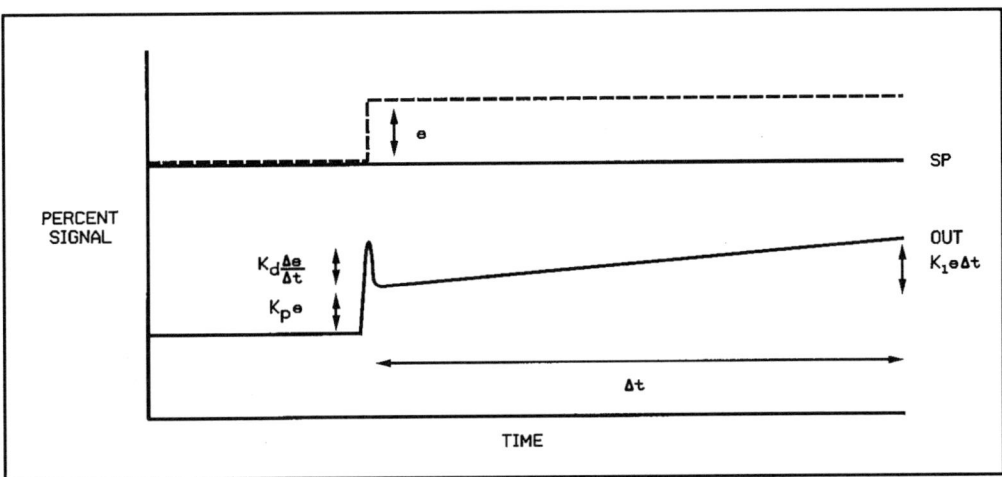

**FIG. 8-5**   Response to step process change—PID control

## PID Control

PID control infers that the controller algorithm consists of proportional, integral, and derivative control.

On the left side of Figure 8-5, before the step change, the process variable and set point are shown to be equal in a PID controller. As such, the output remains at its previous value because no control action is necessary.

When the process variable changes, an error (e) develops, and the output is modified by the proportional, integral, and derivative control algorithms. As shown in Figure 8-5, when the step change occurs, the proportional control causes the output to increase in proportion to the error, while the derivative control causes a spike that is proportional to the rate of change of the error to occur. The integral control algorithm exhibits no immediate response, but causes the output to ramp by an amount that is proportional to the error over time. These actions can be represented by the summation of the proportional, integral, and derivative control algorithms:

$$OUT(t + 1) = OUT(t) + K_p * e + K_i * \Sigma (e * \Delta t) + K_d * \Delta e / \Delta t \tag{8-5}$$

The above equation illustrates that the controller output response to an error is dependent upon the sign and magnitude of the error, and the tuning constants $K_p$, $K_i$, and $K_d$. Controller tuning consists of adjusting these three variables in such a manner that the process variable stays at its set point under process operating conditions. The PID velocity (incremental) algorithm can be expressed as:

$$\Delta OUT = K_p \Delta e + K_i e \Delta t + K_d \frac{\Delta e_t - \Delta e_{t-1}}{\Delta t} \qquad (8\text{-}6)$$

***PID Controller Implementation***   The above discussion, describing the PID controller, illustrates the operation of a parallel PID controller in which each control algorithm is performed independently and subsequently summed. Therefore, there is no interaction between the three tuning parameters; that is, changing the value of one parameter has no effect on the other two.

PID controller implementations utilize control algorithms that create interaction between the tuning parameters to simplify loop tuning. In addition, the algorithms may be performed either in series or in parallel because of electronic or computational limitations. The user should be advised that this can cause different controller output responses from two nonidentical controllers with identical tuning parameter values. In addition, identical controllers can exhibit the same symptoms when they are constructed with components that are subject to drift, such as those components used in mechanical, pneumatic, and nondigital electronic controllers.

## Summary

The PID control algorithm has been the workhorse of industrial process control; its importance to world-wide industrial growth during the last half-century should not be understated. Understanding the operation of this important algorithm and its nuances can be the key to improving control of the process.

## For Further Information

Murrill, Paul, *Fundamentals of Process Control Theory, Second Edition,* ISA, 1991

## ❖ Chapter 9

# *Controller Tuning*

Control loop tuning is the process of verifying the appropriateness, location, installation, and adjustment of *every* device in the control loop, including the equipment, field instrumentation, signal transmitter, controller, and final control element.

Controller tuning, the adjustment of the tuning parameters associated with a controller, is but one facet of control loop tuning. Before controller tuning is attempted, the equipment must be adjusted and operational, the controller input must accurately represent the process variable, and the final control element must be functioning properly from the controller while in manual mode. Failure to perform checks to verify the functioning of these items before tuning the controller can result in poor control and potentially can result in equipment damage and personal injury.

A number of methods can be used to make controller tuning parameter adjustments. A few of these techniques are discussed below to give different perspectives on the subject. Regardless of the method employed, it is important that the person adjusting the controller carefully observe the control loop response for a sufficient time before making controller tuning constant changes.

## Controller Tuning Strategies

Before controller tuning constants are set, it is important to determine the desired strategy of the control loop. Control loop performance can be measured using a number of techniques, such as the percentage overshoot, time to return to set point, integrating the error, and integrating the absolute value of the error. The concepts behind these criteria are valid, but they are primarily of theoretical value in actual plant operations.

The most important determination is whether the control loop should be tuned for set point changes, for process disturbances, or for reduced overshoot. The desire for stability (tuning for process disturbances) and operator adjustment (tuning for set point changes) usually precludes tuning for only one of the above strategies. As such, tuning is usually performed for one strategy, after which the acceptability of the tuning with regard to the other strategy is verified with compromises made as necessary. It should be noted that tuning for an inappropriate response can reduce the effectiveness of the control loop in the process.

***Tuning for Set Point Changes***  Controllers that are tuned for set point changes react quickly to changes to the controller set point. Control loop response can be checked by observing the process variable response to a controller set point change that is made after the process has been stable for a sufficient period of time that the only upset condition is the set point change. The process variable may overshoot the set point and exhibit dampened oscillations before reaching the set point (see Figure 9-1). Control loops that benefit from this tuning strategy are usually not subject to significant disturbances from the process.

For example, a flow loop may benefit from tuning for set point changes when the operator periodically adjusts the flow based on current process conditions. It should be understood that the control loop must still maintain stability and handle minor disturbances from the process. Most flow loops, some temperature loops, and some pressure loops should be tuned for set point changes.

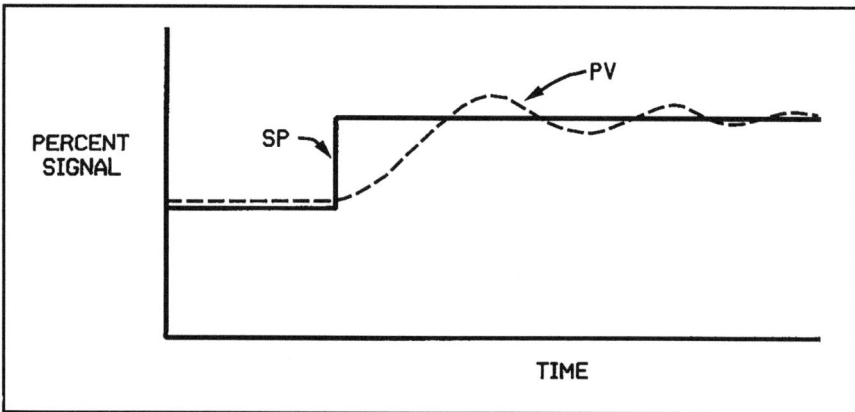

**FIG. 9-1.**   Control loop response (tuned for set point response)

***Tuning for Process Disturbances*** Tuning a controller for disturbances from process upsets involves the examination of control loop operation when process variable upsets occur and no set point changes are made. Control loops that benefit from this tuning strategy do not require set point changes during normal operation, but must more often respond to significant process disturbances. Control loop response can be checked by observing the process variable during the time when process upsets are known to occur (see Figure 9-2).

For example, a boiler drum level controller rarely requires set point changes but must respond to process load changes. Controller response could be checked when a large steam user is turned on (or off) or, perhaps, when starting up. Most level, many temperature, and some pressure loops exhibit similar attributes and should be tuned for process disturbances.

***Tuning for Reduced Overshoot*** Tuning a control loop for reduced overshoot is desirable when a controller output change can cause more significant process upsets than the process variable change that initiated the controller output.

For example, level control of a surge tank is important in order to maintain material in the tank while not overflowing the tank (see Figure 9-3). However, a controller tuned for precise level control may upset other parts of the process by causing abrupt changes in flow. Some loops of all types should be tuned for reduced overshoot.

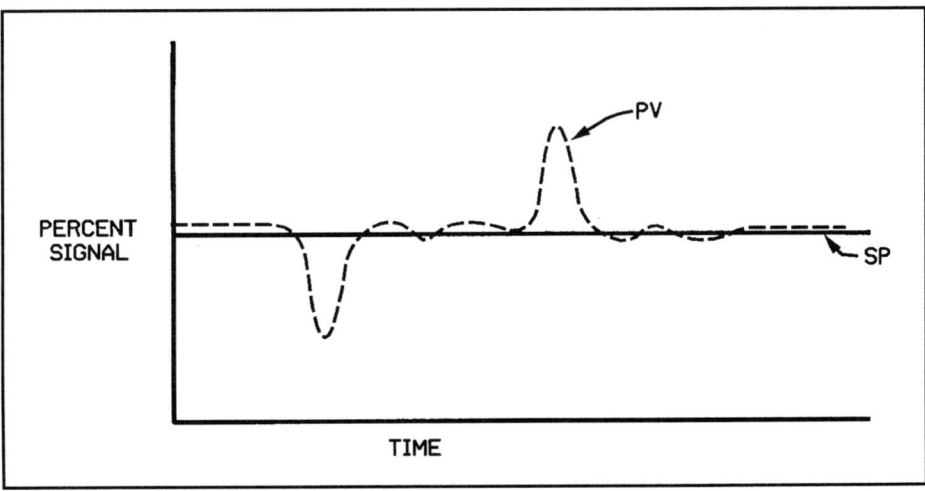

**FIG. 9-2.** Control loop response (tuned for process disturbances)

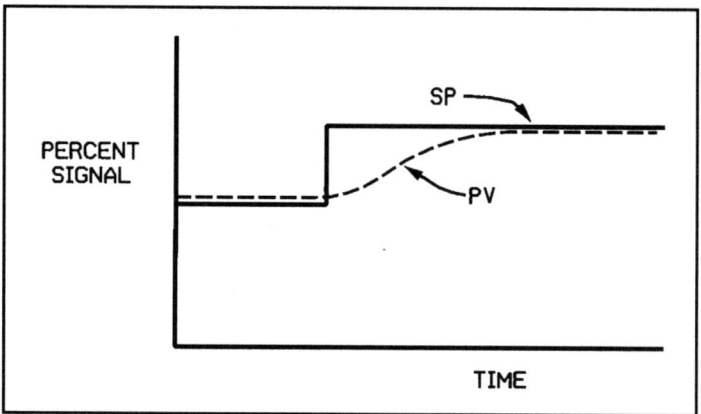

**FIG. 9-3.** Control loop response (tuned for reduced overshoot to setpoint change)

## Quarter Amplitude Oscillation Method

This method is based upon a paper, written in 1942 by J. G. Ziegler and N. B. Nichols, describing a technique by which the tuning constants can be determined. Utilizing this method, the process variable will overshoot and oscillate around the set point with each subsequent oscillation being 25 percent of the amplitude of the previous oscillation.

The closed-loop technique requires that the controller gain be increased in automatic until the loop steadily oscillates about its set point while the integral and derivative algorithms are disabled. The period of oscillation in minutes ($T_u$), while operating in this manner, and the ultimate gain ($K_u$) are used to calculate the proportional, integral, and derivative controller tuning parameters using the following relationships:

|                  | P          | PI            | PID        |
|------------------|------------|---------------|------------|
| $K_p$ (gain)     | 0.5 $K_u$  | 0.45 $K_u$    | 0.6 $K_u$  |
| $T_i$ (minutes)  | —          | $T_u/1.2$     | $T_u/2$    |
| $T_d$ (minutes)  | —          | —             | $T_u/8$    |

The advantage of this technique is that once the period and ultimate gain is known, the controller tuning parameters can be readily calculated. The disadvantage is that the control loop must be allowed to oscillate and may be driven to instability to determine this information, potentially compromising process safety and economic constraints. In addition, this technique can be time consuming and frustrating when the period is long or when variations in the operation of the process impair the ability to obtain accurate data.

Appropriate parameters can also be determined with the loop open by making an abrupt output change and observing the response while the control loop is in manual. The dead time (the time during which the measurement does not react after manipulating the final control element) and the time constant (the time required for the measurement to change 63.2 percent of its response to a step manipulation of the final control element) can be used to calculate tuning parameters for quarter amplitude decay using the following open-loop relationships.

|  | *P* | *PI* | *PID* |
|---|---|---|---|
| $K_p$ (gain) | 1TC/(K*DT) | 0.9TC/(K*DT) | 1.2TC/(K*DT) |
| $T_i$ (minutes) | — | 3.33DT | 2.0DT |
| $T_d$ (minutes) | — | — | 0.5DT |

where TC is the process time constant (minutes), DT is the process dead time (minutes), and K is the open-loop process gain (the ratio of the change in process variable to the controller output change that caused it).

***Alternate Method*** This method, described by G.K. McMillan, is similar to the Ziegler-Nichols open-loop method but does not drive the control loop into oscillation. The integral and derivative algorithms are disabled, and the gain is increased until each successive oscillation peak in the response to a step set point change is 25 percent of the previous peak. The gain ($K_g$) and period ($T_g$)(minutes) of the closed-loop oscillations can be determined and used to calculate the tuning parameters:

|  | *P* | *PI* | *PID* |
|---|---|---|---|
| $K_p$* | $K_g/_{1.2}$ | $K_g/_{2.4}$ | $K_g/_{1.2}$ |
| $K_i$ | — | $T_g$ | $T_g/_2$** |
| $K_d$ | — | — | $T_g/_{10}$** |

  * Multiply by 0.8 for slow loops
 ** Double for loops with dead band (from valve hysteresis)

# Trial and Error Method

A survey of field practice would show this method to be much more commonly used than any of the above methods. There is no formula or formal methodology per se, because each individual has a personal "proprietary" method. However, most of these methods include increasing the gain with proportional control only until the loop is unstable, then increasing the integral. Subsequent responses are observed with adjustments made by "feel."

As noted above, the controller may be tuned for process disturbances, set point changes, or reduced overshoot. When using the trial and error method, it is important to "challenge" the control loop with appropriate upset conditions and tune the controller to produce the desired response.

# Self-Tuning Method

Some controllers contain algorithms that enable the controller to calculate controller tuning constants based upon natural or, more commonly, induced process perturbations.

Some algorithms continually update the controller tuning constants so that the current controller tuning will reflect process response changes. Process response changes may be caused by such things as mechanical sluggishness, mechanical wear, aging of parts, and the like.

Other algorithms calculate suggested controller tuning constants that the user must enter. The approach of maintaining fixed controller tuning constants is based on the premise that the process dynamics remain fixed; fixed controller tuning constants are satisfactory, provided that the proper controller tuning constants are selected. This approach also minimizes the risk of the presence of inappropriate controller tuning constants due to process upsets and interaction.

# Typical Controller Settings

Due to the uniqueness of each process, there are no universal tuning constants, however, G.K. McMillan suggests the following typical controller settings:

| | Proportional (gain) | (% PB) | Integral (repeats/minute) | Derivative (minutes) |
|---|---|---|---|---|
| Flow | 0.2-1.0 | 100-500 | 10-50 | none |
| Liquid Pressure | 0.2-1.0 | 100-500 | 10-50 | none |
| Gas Pressure | 2-100 | 1.0-50 | 0.02-1.0 | 0.02-0.10 |
| Level | 2-100 | 1.0-50 | 0.01-0.10 | 0.01-0.05 |
| Temperature | 1.0-10 | 10-100 | 0.02-0.05 | 0.5-2.0 |
| Chromatograph | 0.125-0.50 | 200-800 | 0.01-0.10 | none |

## Summary

Controller tuning compensates for the dynamics of the control loop, enabling a process parameter to be stabilized in the face of potentially destabilizing factors. To improve control of the process, elimination of potentially destabilizing factors that may be part of the control loop (such as an excessively dampened measurement, an inaccurate instrument, a sluggish control valve, and the like) should be performed prior to controller tuning.

While there are guidelines for tuning controllers, controller tuning is not an exact science and, as such, may require many trial and error adjustments before the control loop functions satisfactorily, especially when controllers are applied to control loops that have long time constants and long response times.

## For Further Information

Corripio, Armando B., *Tuning of Industrial Control Systems* ISA, 1990.

Mollenkamp, Robert A, *Introduction to Automatic Process Control,* ISA, 1984.

McMillan, Gregory K., *Tuning and Control Loop Performance, Second Editon,* ISA, 1990.

St. Clair, David W., *"Controller Tuning and Control Loop Performance,"* Straight-Line Control Company, 1989.

Liptak, Bela G., (Editor-in-chief), *Instrument Engineers' Handbook,* Chilton Book Company, 1985.

## ❖ Chapter 10

# *Regulatory Control Loop Pairing*

Implementing each individual facet of instrumentation control loop design correctly is critical to the proper operation of the control loop. The field measurement device must accurately measure the process variable, and the final control element (typically a control valve) must be capable of adequately manipulating the process. The controller should contain the control algorithms and features to control the process. All devices in the loop must be properly interconnected and calibrated.

However, implementing each individual facet of instrumentation control loop design correctly is no guarantee that the control loop will function properly in the actual process, even if the control loop is operating as designed.

Previous chapters have focused on the function of each individual part of a control loop and the importance of implementing each function correctly. This chapter focuses on the need to select the proper process variable to control and the proper final control element to manipulate.

## Regulatory Control Strategy

A regulatory control loop is defined herein as a control loop that contains one measurement device, one controller, and one final control element. Flow, level, pressure, and temperature measurements comprise the majority of industrial measurements. PID controllers dominate the industrial controllers, while the majority of final control elements used in industry are control valves.

Proper operation of the control loop requires that each part of the control loop function properly, both individually and as a loop, and the field equipment must be properly connected to the process.

In addition, the control loop must implement the proper control strategy. For regulatory control loops, the proper process variable is selected to feed the controller that is used to manipulate the proper final control element. This selection process is called pairing.

In practice, determining the regulatory control loop strategy and pairing the process variables with final control elements is usually based on experience. It is *essential* that the person determining the control strategy have intimate knowledge of the process. It should be remembered that THE GREAT MAJORITY OF FINAL CONTROL ELEMENTS MANIPULATE FLOW, even though it may be desirable to control a process variable other than flow. Further, the manipulated, final control element should be capable of affecting the process variable that is to be controlled with an appropriate installed charecteristc.

Also, manipulation of what appears to be the appropriate final control element in order to control the desired variable may have an undesirable effect on another part of the process. When this occurs, more sophisticated techniques may be required to properly control the process.

***Flow Control***  The most straightforward application of regulatory control is that of controlling flow, where flow is the only process variable of concern.

In the application shown in Figure 10-1, the flow leaving a tank is controlled by measuring the flow and manipulating the

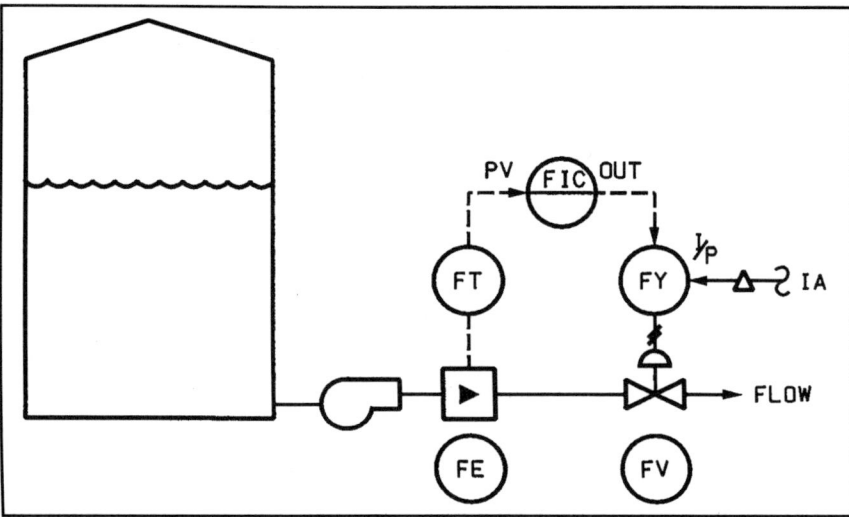

**FIG. 10-1.A**  Regulatory control of flow. Regulatory control loop with control valve.

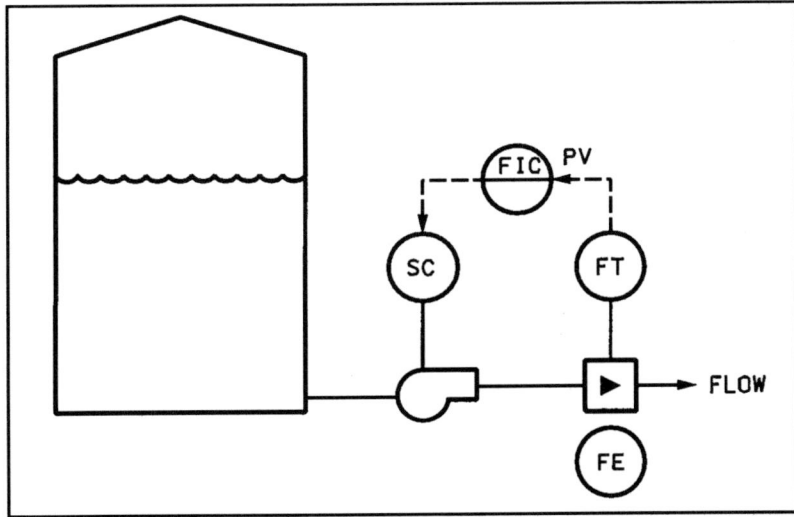

**FIG. 10-1.B**   Regulatory control of flow. Regulatory control loop
with variable-speed drive.

valve in line with the flowmeter or the variable-speed drive
rotating the pump in series with the flowmeter. In the case
of the control valve, should the flow increase (decrease), the
flow controller will throttle the valve more closed (open) to
bring the flow down (up) to the set point. If the flow were
to be paired with another valve in the process, the desired
flow would not be under control.

***Level Control***  If it were desirable to control the level in a tank
because the tank could overflow or its level could become too
low to maintain pump suction, a level controller could manipulate a makeup valve or a discharge control valve.

Selecting the proper valve for manipulation depends on the
process and the effects of manipulation on the upstream
and/or downstream operations (see Figure 10-2). Note that an
increase (decrease) in tank level will cause the level controller
to throttle the makeup valve more closed (open) to bring the
level down (up) to its setpoint by manipulating the flow to the
tank. Similarly, the level controller that controls the discharge
valve will throttle the discharge valve open (closed) to bring
the level down (up) to its set point by manipulating the flow
out of the tank. If the tank level were paired with the wrong
valve, a process upset could occur elsewhere in the process,
even though the level control loop was functioning properly.

In many installations, flowmeters are installed to monitor
flows that seem to be important.

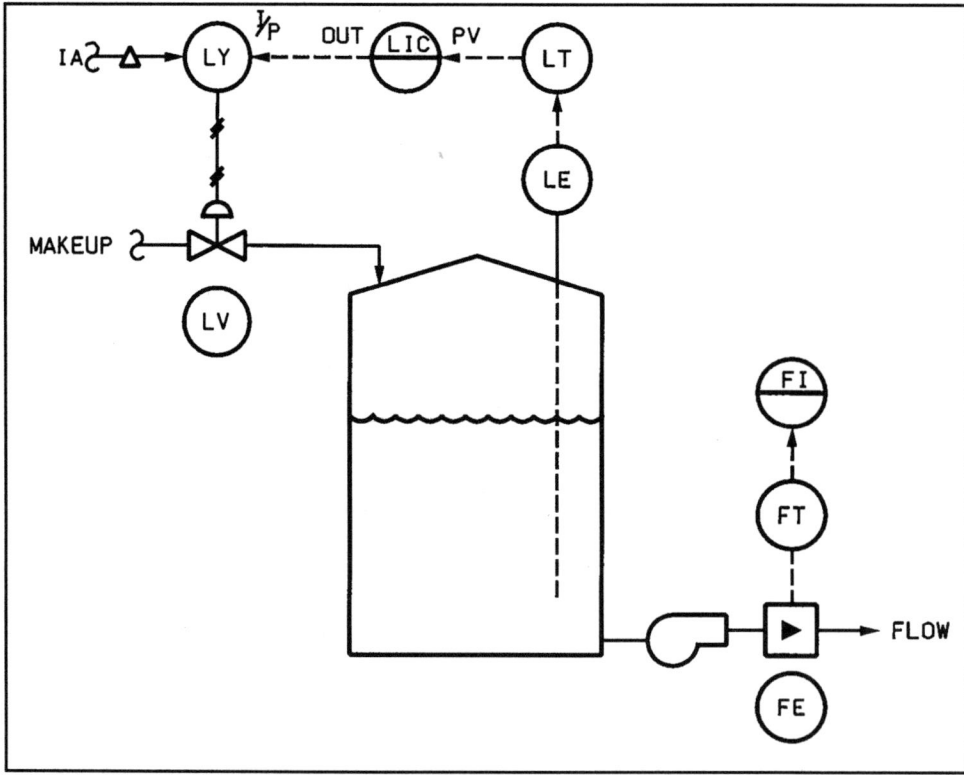

**FIG. 10-2.A**   Regulatory control of level. Regulatory level control loop controlling inlet flow.

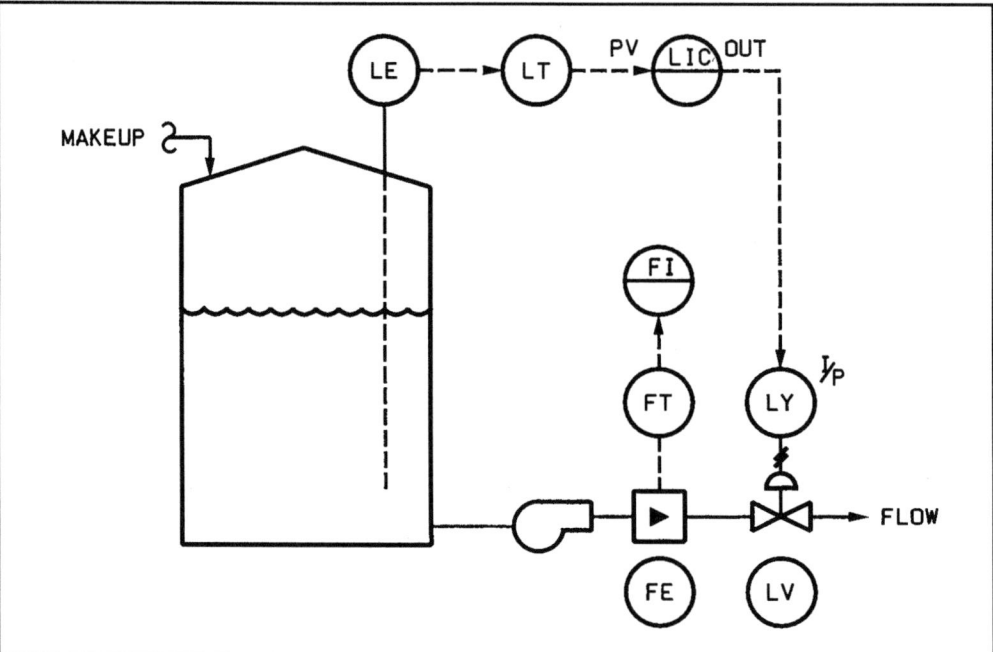

**FIG. 10-2.B**   Regulatory control of level. Regulatory level control loop controlling discharge flow.

**Pressure Control**  If it were desirable to control the pressure in a vessel, a pressure controller could manipulate the inlet valve or the discharge control valve (see Figure 10-3).

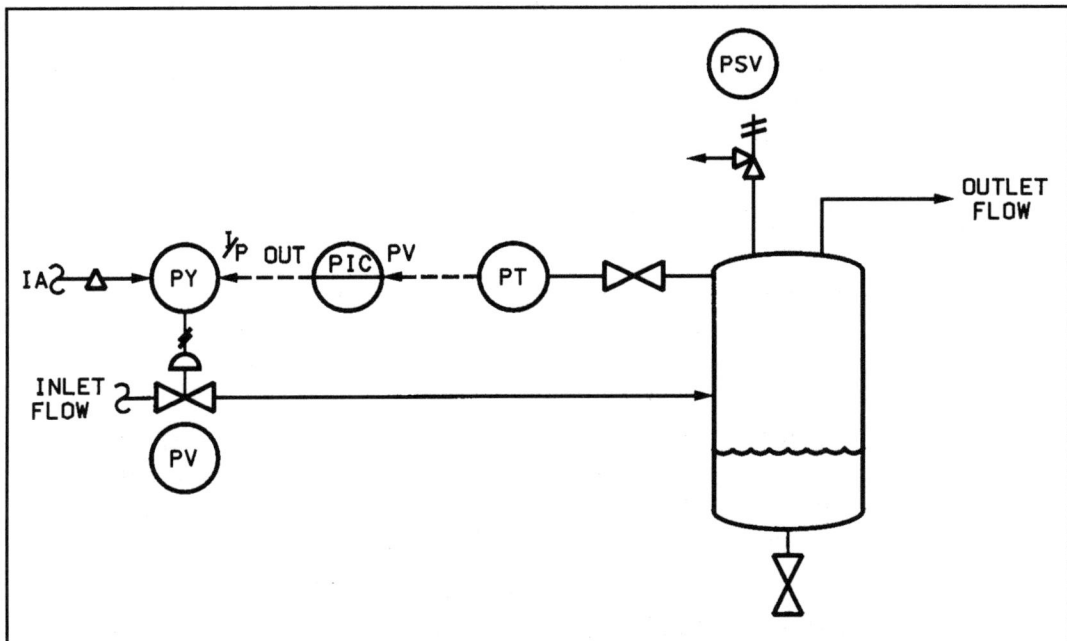

**FIG. 10-3.A**   Regulatory control of pressure. Regulatory gas pressure control loop controlling inlet flow.

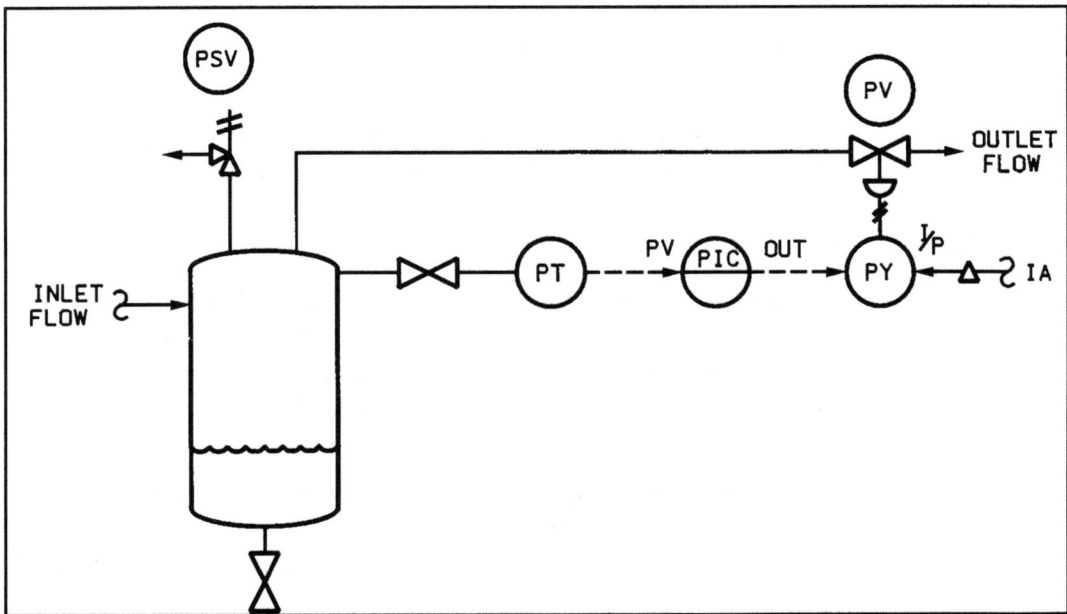

**FIG. 10-3.B**   Regulatory control of pressure. Regulatory gas pressure control loop controlling discharge flow.

Selecting the proper valve for manipulation depends upon the process and the effects of manipulation on the upstream and/or downstream operations. Note that an increase (decrease) in tank pressure will cause the pressure controller to throttle the inlet valve more closed (open) to bring the pressure down (up) to its set point by manipulating the flow to the tank. Similarly, the pressure controller that controls the discharge valve will throttle the discharge valve open (closed) to bring the pressure down (up) to its set point by manipulating the flow out of the vessel. If the pressure were paired with a valve that did not affect the pressure, the pressure control loop would not function properly.

Flowmeters are usually not installed to monitor flows in pressure control applications unless the flows are unusually important.

***Temperature Control*** If it were desirable to control the temperature of the discharge flow of a heat exchanger, a temperature controller could manipulate the heating valve or the cooling control valve (see Figure 10-4).

Note that an increase (decrease) in the discharge temperature will cause the heating temperature controller to throttle the steam valve more closed (open) to bring the temperature down (up) to its set point by manipulating the steam flow to the heat exchanger. Similarly, the temperature controller that controls the cooling water valve will throttle the discharge valve open (closed) to bring the temperature down (up) to its set point by manipulating the cooling water flow.

Flowmeters are usually not installed to monitor utility flows unless the flows are large enough to be significant.

## Summary

Regulatory control (as defined above) is a collection of individual control loops that are limited to one process variable and one manipulated variable. Most regulatory control loops manipulate flow to control the desired process variable.

Should the control strategy and pairing be inappropriate, it may not be possible to bring the process under control. Further, seemingly appropriate strategies and pairings may upset the process.

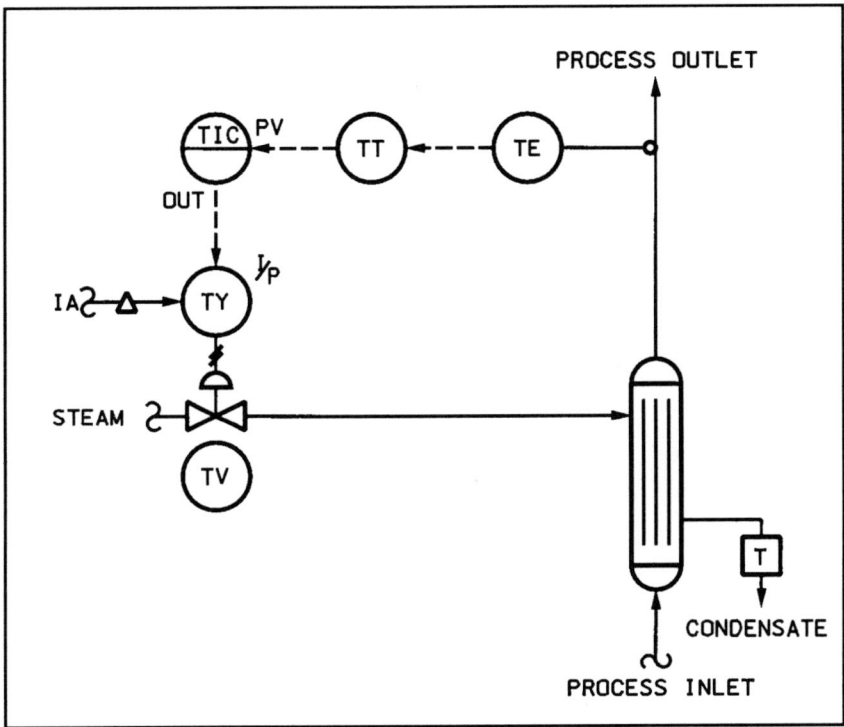

**FIG. 10-4.A** Regulatory control of temperature. Regulatory temperature control loop (heating with steam).

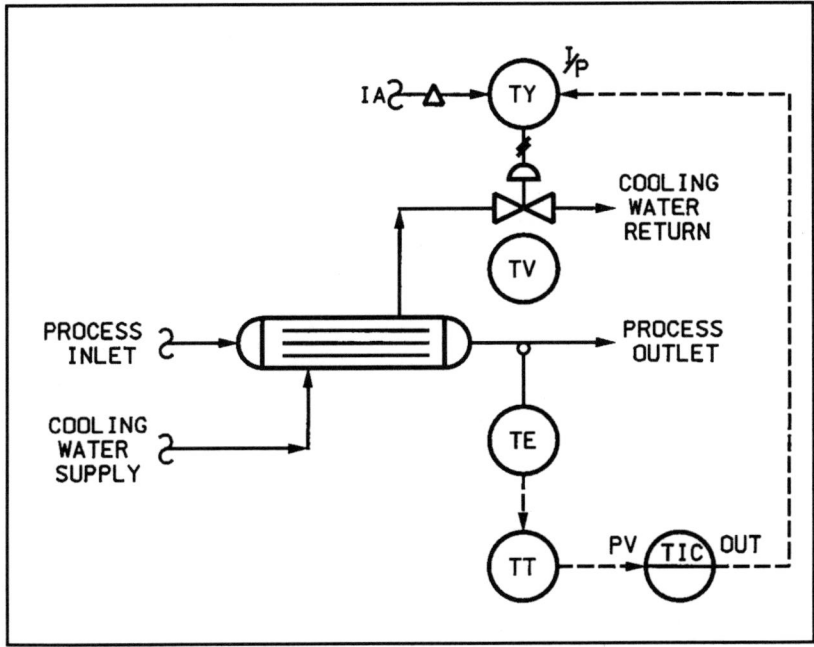

**FIG. 10-4.B** Regulatory control of temperature. Regulatory temperature control loop (cooling water).

## ❖ Chapter 11

# *The Limitations of Regulatory Control*

Regulatory control has been available for more than 50 years, and its single input-single output nature allows it to be easily understood and applied with apparent success. This success is usually evaluated using such criteria as: (1) the flow was not excessive; (2) the tank did not overflow; (3) the pressure relief valve did not operate; and (4) the product did not boil or freeze.

While these expectations may have been reasonable decades ago, they are unrealistically low given the control tools available. It should be noted that the relative simplicity of regulatory control and its criteria for success have created a large cadre of people with low expectations and the belief that regulatory control is all that will ever be required.

Having presented and described regulatory control in the pristine environment of steady-state operation that represents reality to so many, the problems associated with some regulatory control loops under dynamic conditions can be explored.

## Control Loop Response

Control loops can be roughly divided into loops that respond relatively quickly (fast loops) and those loops that respond sluggishly (slow loops).

Most flow control loops are fast loops: when the control valve (or variable speed drive) is manipulated, the flowmeter usually senses the change almost immediately. Some gas flow control loops can have dampened responses due to long pipe lengths, fluctuating gas pressure, large pipe size, and flowmeter technology. Despite this sluggishness, even these flow loops are usually the fastest loops in the process and can be considered "fast." See Figure 11-1.

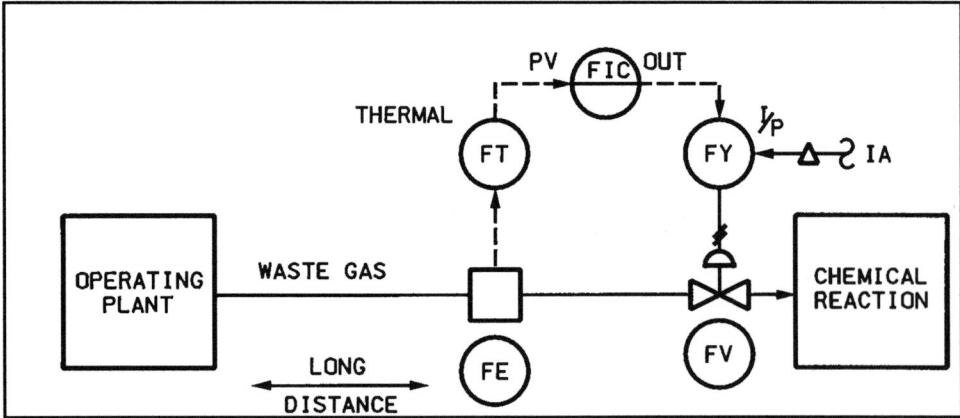

**FIG. 11-1.**    Flow control loop with fast response dampened by pipe length and flowmeter technology

Most pressure control loops are also fast loops, especially where the controlled volume and flow are small compared to the control valve capacity. In most applications, pressure loops are slower than the flow loops, but react quickly enough to be considered fast. See Figure 11-2.

Pressure loops can be considerably slower when the control valve capacity is small compared to the volume and to process flow.

Level control loops are typically slow loops because deviation from set point usually causes a small change in flow relative to

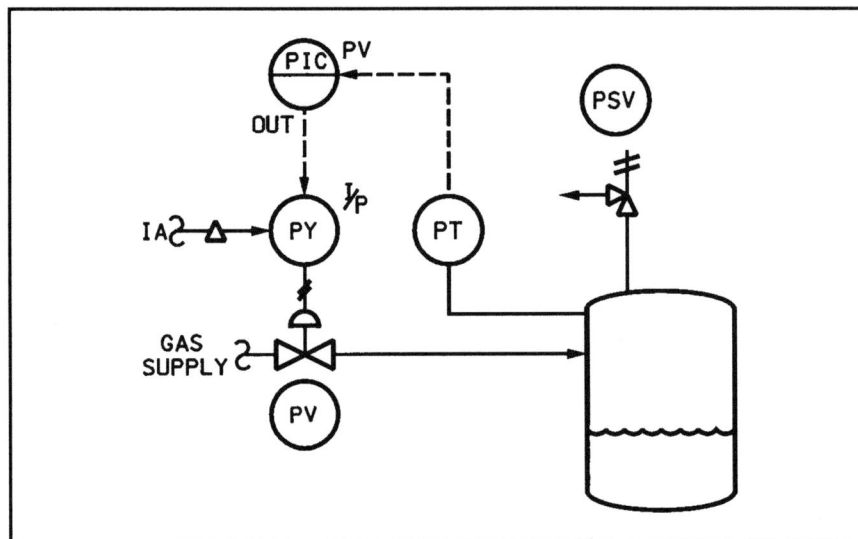

**FIG. 11-2.**    Gas pressure control loop

tank volume. It is not uncommon for properly tuned level loops to require tens of minutes to stabilize. The exception is the case in which the control valve capacity is large compared to the tank volume, a situation that is indicative of an undersized tank, an oversized control valve, an unnecessary tank, or an inappropriate control strategy.

Temperature control loops are usually slow loops, especially when a heat exchanger is involved. See Figure 11-3. It is not uncommon for heat exchangers to exhibit process dead times of tens of minutes, i.e., the time necessary for the process variable measurement (process discharge temperature) to start sensing the effect of a heating or cooling medium change. In addition, the process time constant (that is, the time necessary for the process variable to change 63.2 percent of its change after the process dead time) can be in minutes for heat exchangers of significant size relative to flow.

Some processes, such as furnace temperature control, can have shorter dead times and time constants and exhibit a significantly faster response than described above for heat exchangers. Despite this improvement, temperature loops are

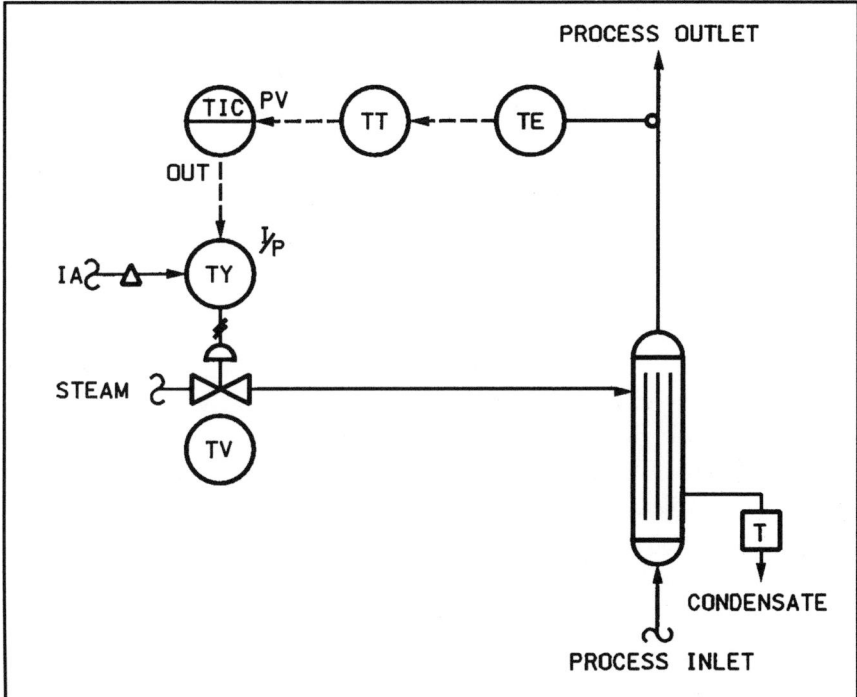

**FIG. 11-3.** Regulatory temperature control loop (heating with steam)

usually sufficiently slow (compared to other control loops) in the process that they can be considered slow loops.

***Fast and Slow Loop Tuning***     Barring uncommon idiosyncracies, fast control loops can be tuned satisfactorily using the proportional and integral control algorithms, because the process variable responds to final control element changes almost immediately. As a result, the effect of controller tuning adjustments can be observed in seconds or minutes, and prompt additional controller tuning adjustments can be likewise analyzed.

Slow control loops can represent a formidable tuning challenge. Consider regulatory temperature control of the process output of a heat exchanger using steam as the heating medium (see Figure 11-3).

The apparent control strategy is to control the process fluid temperature at a desired temperature. As such, the heat exchanger process outlet temperature is measured and controlled by manipulating the steam valve. Using this pairing, it would seem reasonable that, if the temperature increased (decreased), the controller would sense the deviation and throttle the steam valve more closed (open) to maintain the temperature at the desired set point. This reasoning implicitly applies to the steady state, that is, the steam valve will be more closed (open) when the temperature is returned to set point. However, this approach completely ignores the actual control loop dynamics.

Assuming that the temperature is stabilized at its set point, changing the set point will move the steam valve, but will not affect the measurement during its dead time (of minutes) to allow time for heat transfer and time for the heated fluid to reach the location of the temperature measurement (see Figure 11-4). During the dead time, the lack of process variable response maintains the deviation between the set point and process variable, which causes the integral controller algorithm to continue to move the steam valve in what appears to be the proper direction. When the heated fluid reaches the measurement location, the measurement changes relatively rapidly, causing the steam valve to relatively abruptly move in the opposite direction.

Because a properly tuned temperature controller may take tens of minutes or even hours to stabilize, it is not uncommon to make tuning changes twice a day—once when arriving

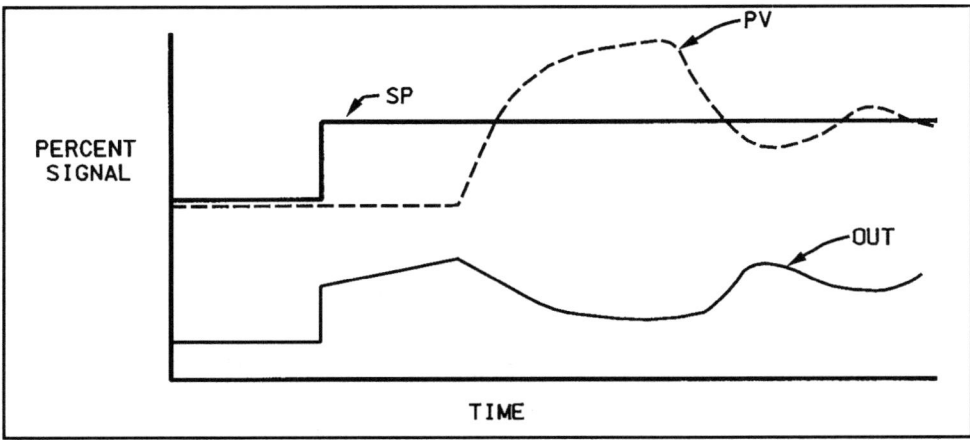

**FIG. 11-4.** Regulatory temperature control loop response

at work and again prior to leaving work. To complicate the is-
sue, in addition to the proportional and integral algorithms
used for fast loops, the derivative algorithm must be used in
many applications. In order to maintain stability, these loops
are usually slow and sluggish to react to either set point or
process disturbances. The difficulty and frustration of tuning
these loops and the poor dynamic loop performance are usually not recognized by anyone except the individual performing the tuning.

## Process Upsets

Virtually all regulatory control loops manipulate a process
flow. Set point changes and process disturbances to slow loops
may cause flow changes that have a small effect on the process
variable but a large effect on the operating process.

The bottoms level of distillation column A (see Figure 11-5) is
controlled by manipulating the bottoms control valve that
throttles the flow of liquid leaving the column to directly feed
distillation column B. In this application, the set point is
rarely changed, so the control loop should be tuned for
process disturbances that can and will occur. Because small
changes in the control valve position have a negligible effect
on the level, the level control loop is often tuned in such a
manner that relatively small level disturbances will cause significant movement of the control valve, in turn causing relatively large flow fluctuations, even in so-called steady state.
These fluctuations can have a significant effect on the operation of distillation column B, even though the level loop is
functioning properly.

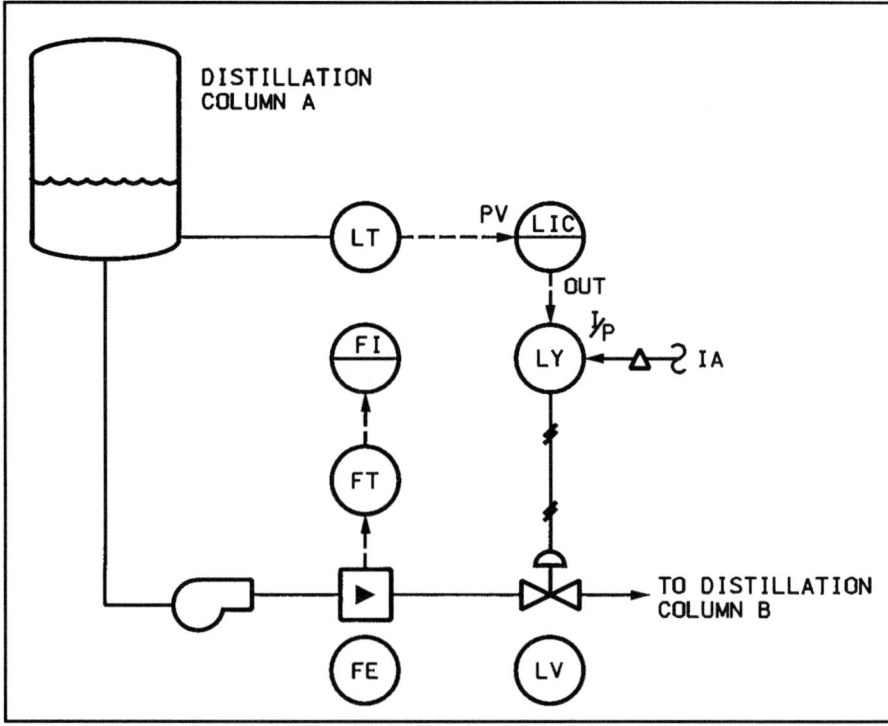

**FIG. 11-5.**    Regulatory temperature control loops with interaction

## Control Loop Interaction

Few processes are a collection of totally independent entities that can be satisfactorily controlled by a number of regulatory control loops with one measurement, one controller, and one final control element. As illustrated in the level control loop above, the operation of one control loop (level) may have a pronounced effect another (flow).

As a result of the cooling water supply and return headers being undersized, the pressure difference between the supply and return headers will vary with the cooling water flow rate in the control loop shown in Figure 11-6. Increasing the cooling water flow in Process A will cause the cooling water flow through the Process B heat exchanger to be reduced. The effects of this reduction will not be measured until some time later, after which the Process B heat exchanger cooling water will be increased, in turn reducing the cooling water to Process A, and so on. In this application, what appear to be unrelated regulatory control loops on different processes can interact with one another to cause process upsets.

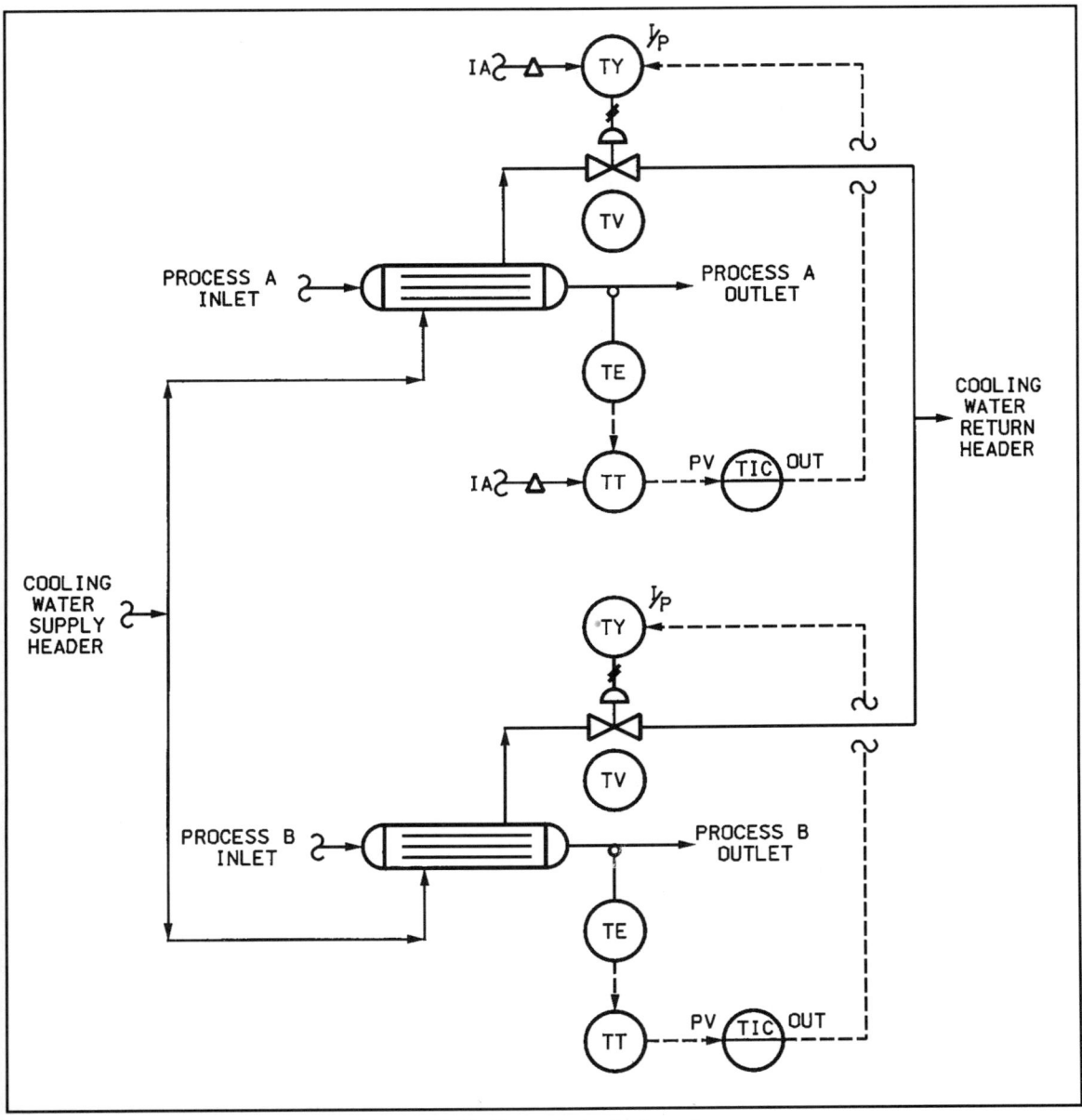

**FIG. 11-6.** Regulatory level control loop (distillation column bottoms)

## Summary

It is not uncommon to be exposed to lengthy analyses about how each regulatory control loop stabilizes its respective process variable while implying (but not stating) that the process is at steady state. Real processes exhibit dynamic considerations that are absent from such analyses. How well these considerations are compensated for by the control system often determines the quality of control and, in some applications, the ability to control. Of significant concern is whenslow loops *measure* process variables that react slowly but *manipulate* process variables that react quickly.

The ability of a regulatory control loop to compensate for process dynamics, process upsets, and control loop interaction are limited by the single process variable and single manipulated variable nature of the regulatory control loop. It is estimated that over 50 percent of existing regulatory control loops could benefit from some form of more sophisticated control.

# ❖ Chapter 12

# *Advanced Regulatory Control Tools*

Advanced regulatory control tools include mathematical functions that are used in conjunction with feedback, feedforward, cascade control, feedback trim, dynamic compensation, and/or feedforward action to implement advanced regulatory control strategies. The intent of these more sophisticated control strategies is to manipulate the final control element to obtain superior control more intelligently than would otherwise be possible using regulatory control.

Some of these concepts are not new and have been around for decades, but the availability of digital controllers and distributed control systems have reduced their cost and simplified their implementation. However, other tools have become more sophisticated, for example, the controller set point may track the process variable and freeze its output to avoid upsets when a bad measurement is detected.

## Mathematical Functions

Mathematical functions represent a set of very powerful tools that can be used to significantly improve control of the process when applying advanced regulatory control strategies. Some of the typical tools that are available in digital controllers (in limited quantities) and distributed control systems are reviewed below.

*Calculation* Calculation capabilities may be limited to preset formulas, but it usually is possible to write algebraic equations that will perform calculations using real-time measurements. Available functions typically include addition, subtraction, multiplication, division, ratios, exponentials, logarithms, trigonometric functions, and the like.

***Characterization*** A characterizer defines a fixed relationship between its input signal and its output signal that is usually defined from tables of corresponding input and output data from which the characterizer extrapolates the output signal corresponding to a given input. Linearization of an input signal from a look-up table is a typical application of the characterization function.

***Dead Time Algorithm*** The dead time algorithm is a dynamic compensator that is used to delay a signal by a specified period of time and is primarily used where the effects of the signal on the process would be premature and, as such, must be delayed (see Figure 12-1).

Dead time is often applied in the feedforward signal path in conjunction with a lead/lag algorithm and in processes that contain transportation time delays.

***Lead/Lag Algorithm*** The lead/lag algorithm is a dynamic compensator that is almost exclusively used in the feedforward signal path. The lead algorithm is used to immediately affect the process as a result of a feedforward variable change. The lag algorithm is used to dampen the effect of a signal on the process. (See Figure 12-2).

A combination of lead and lag can be applied to a signal to obtain an immediate response followed by a damp-ened response over time. In some applications, failure to use a lead/lag algorithm to dampen the response can trigger process upsets, as can maladjustment of the lead/lag algorithm.

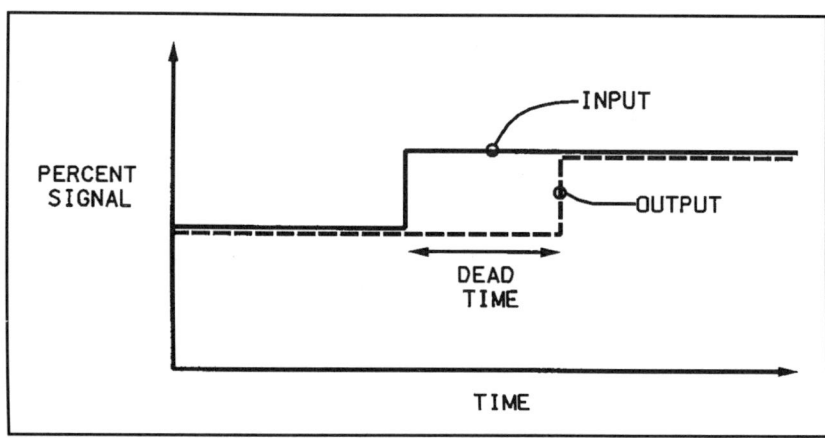

**FIG. 12-1** Dead time.

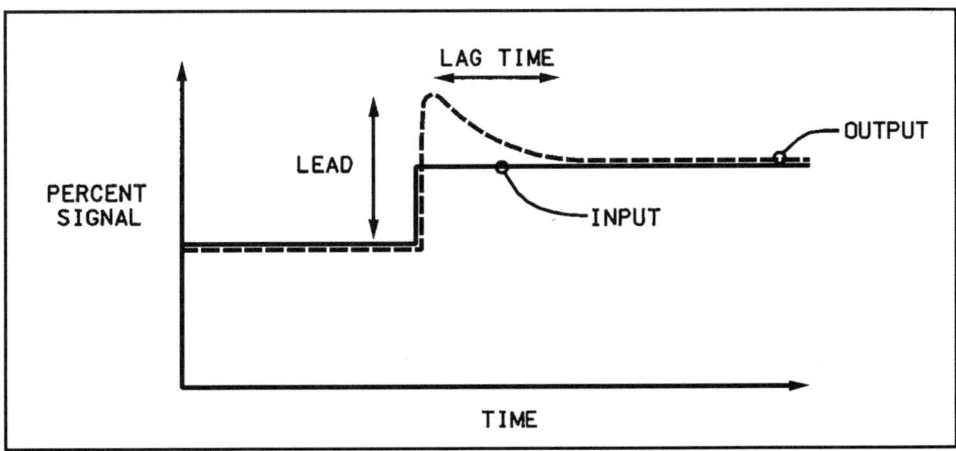

**FIG. 12-2.**   Lead/lag algorithm

***Logic and Data Manipulation***  Utilizing field measurements, control systems can determine when a given process condition exists and can manipulate data accordingly. For example, when the output of one controller is at its maximum, another controller could be activated to control another process variable in automatic operation with a preset set point. The ability of some distributed control systems to perform logic and manipulate data is virtually limitless.

***Ramp Function***  The ramp function is used to increase (decrease) a signal at a constant rate between its starting and ending points. This function may be desirable when used to minimize abrupt changes that may upset the process, such as slowly opening a control valve to prevent damage from a large flow increase.

***Signal Limiter***  Signal limiters are used to clamp a signal between high and low limits.

***Signal Selection***  The function of a signal selector is to select one input (as its output) based upon its signal selection criteria, such as high signal, low signal, average signal, and the like. This function is often used in combustion controls to select the highest (lowest) of the desired and required air (gas) flows to ensure sufficient air to combust the fuel.

## Advanced Regulatory Control Process Variables

Mathematical tools available within the realm of advanced regulatory control can be used to calculate process variables that are more accurate than the raw measurement and more representative of the process. Based upon a number of process measurements, it is possible to calculate process variables that have physical significance but can not be measured directly and may not physically exist. In addition, mathematical functions can be used to correct for processes that are mathematically understood.

Determining which process variables are of importance and should be controlled in a given application requires an intimate knowledge of the process, an understanding of the appropriate measurement techniques, and experience. It is important to get as close to the process as possible, measure what is to be controlled, understand what is being measured, and implement those measurements with the understanding that the measurement does not stop at the transmitter. Both assessing the reliability of the selected instruments and the chances of success are based on experience.

***Flow Measurement***   Flow measurement presents a number of challenges that are usually not addressed, such as the type of flow measurement required for a specific application. Flowmeters are available to measure mass, volume, velocity, and dynamic head, not all of which may be suitable for the process at hand.

If it is desired to measure and control the mass flow of a liquid with varying specific gravity, it would seem logical that a mass flowmeter should be used. However, it is not uncommon to find a flowmeter that measures dynamic head scaled to mass units (which it does not measure). Moreover, this measurement is misleading and does not accurately represent the process.

When a flowmeter is not available to measure the desired type of flow, it may still be possible to improve the measurement by mathematically compensating for fluid properties. The equations for the installed flowmeter should be obtained and examined to determine how the fluid properties affect the flow measurement. An equation can then be mathematically implemented to correct the raw flow measurement for the fluid properties (which are measurable or available) to obtain a more accurate flow measurement.

***Pressure and Temperature Compensation***   The effects of pressure and temperature are far more pronounced on gas flow measurements than on liquid flow measurements.

Nonetheless, the number of gas flow control loops that do not compensate for pressure and temperature is astounding. For example, at ambient temperatures, a 3-degree Celsius gas temperature change affects the volumetric flow by approximately 1 percent, while a pressure change of 0.3 bar (4.5 psi) at 2.0 bar (29 0 psi) will affect volumetric flow by approximately 10 percent. If the raw flow measurement is not corrected for these and other process variations, the instrumentation will not be controlling the proper parameter, despite the apparent control of the (incorrect) raw flow measurement signal. This scenario, and worse, is not uncommon in actual processes where maintaining a constant known flow is extremely important to the safe and economical operation of the process.

In some applications, the Ideal Gas Law can be used to correct the raw flow signal by multiplying the raw volumetric flow signal by

$$(P \times T_{ref}) / (P_{ref} \times T) \qquad \text{(12-1) (velocity flowmeters)}$$

$$\sqrt{(P \times T_{ref}) / (P_{ref} \times T)} \qquad \text{(12-2) (head flowmeters)}$$

where P and T are the absolute pressure and temperature, respectively, and the subscript $_{ref}$ denotes the reference conditions to which the flowmeter was calibrated. When available, more definitive process and fluid information should be used.

***Level Measurement***  Most level measurement devices measure either actual level or the head of liquid at a known tank location. Variations in liquid composition and temperature can cause specific gravity changes that can significantly affect the accuracy of head measurement. With sufficient process information, an equation can be written to compensate for the specific gravity changes. However, putting another liquid with a different specific gravity in the vessel can cause significant measurement error when the transmitter is not recalibrated or when the calculation is not altered.

Fortunately, most industrial level control loops need not control the level very precisely, although the operation of some processes may be improved by doing so.

***Pressure Measurement***  Process pressures that can be measured include absolute pressure, gage pressure, and differential pressure. The application of each type of transmitter may seem obvious, but there are applications in which this is not the case.

For example, it would seem that the differential pressure across an atmospheric distillation column would be measured with a differential pressure transmitter, but condensation can occur in the impulse lines, which can affect the measurement. Using capillary tubing with diaphragm seals would resolve this problem, but the taps may be so far apart that the maximum capillary length is exceeded. Two gage pressure transmitters could be applied and their signals subtracted to obtain the differential pressure, while making the pressure information available. However, for chemical reasons, the absolute column pressure is of interest, and because atmospheric pressure changes will not be measured by the gage pressure transmitter, two absolute pressure transmitters should be applied. In addition, these transmitters must be accurate enough to correctly measure the differential pressure.

**Temperature Measurement**  Temperature measurements can be measured by a number of devices, such as thermocouples, resistance temperature detector (RTDs), infrared pyrometers, etc. It is important that the temperature-sensing technique used be appropriate for the application and *properly located*. In addition, appropriate signal conditioning should be performed to achieve accurate temperature measurement in the range of interest.

**Composition Measurement**  Many techniques can be used to measure composition, but, regardless of the technique, the analyzer should sample the process at the correct location. In addition, when the analyzer is not *in situ,* it should be close enough to the process to avoid exhibiting excessive transport delay, which will add to the analysis time.

**Heat Flow**  Many processes would benefit from the control of heat flow, a process variable that cannot be measured but can be calculated using appropriate process measurements.

In a multifuel furnace in which the waste fuel flows vary, it is common to control furnace temperature by adjusting the purchased fuel through the use of a temperature controller. Superior control could be obtained by stabilizing the total amount of fuel that enters the furnace. The total fuel entering the furnace can be calculated by summing each fuel flow multiplied by its respective heating value, which can be measured or estimated:

Total Fuel Content $= (Q_1 * HV_1) + (Q_2 * HV_2) + \ldots$     (12-3)

where $Q_n$ is the nth fuel flow and $HV_n$ is the heating value of the nth fuel in like units.

In a flow stream, such as would flow through a heat exchanger, the heat gained or lost can be calculated by the following:

Heat Gain (Loss) $= m * c_p * \Delta t$     (12-4)

where m is the fluid mass flow, $c_p$ is the specific heat of the fluid, and delta t is the temperature difference between the fluid inlet and fluid outlet. Be aware that process dynamics, especially prevalent in equipment exhibiting a long dead time and time constant, such as heat exchangers, can alter the accuracy of this calculation during transients.

# Feedback Control

A feedback controller compares the process variable (measured in real time) with the set point and manipulates the final control element based upon the controller tuning adjustments. This type of feedback control, in the absence of any other control strategy, is regulatory control. Regulatory control and its limitations have been discussed at length elsewhere and will not be discussed here. It should be noted, however, that virtually all control loops utilize some form of feedback control in their control strategies.

***Feedback Control with Variable Gain***  In some applications, the process variable changes nonlinearly with the manipulated variable.

For example, when controlling pH by manipulating reagent flow, the effect of increasing reagent flow by large amounts when the pH is relatively low or high is a small pH change (see Figure 12-3). When the pH is near 7, small changes in reagent flow can drastically affect the pH. Were a PID controller applied to this process, the reagent flow would change based upon the error, regardless of whether or not the pH was near 7. Therefore, tuning the controller near (far from) a pH of 7 would result in an extremely slow (fast) pH response when the pH is far from (near) a pH of 7.

To compensate for the changing process gain, some proprietary controllers allow the characterization of the gain, so that

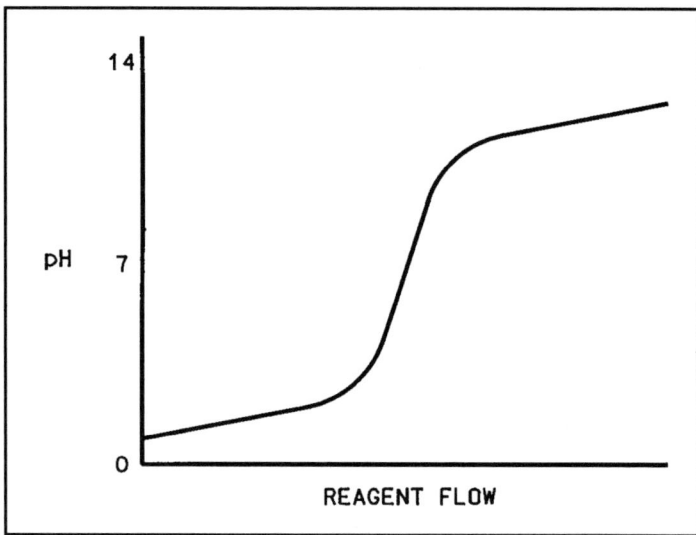

**FIG. 12-3.**    Relationship between reagent flow and pH

the pH controller makes substantial reagent adjustments when the pH is far from 7 and small reagent adjustments when the pH is near 7. Other proprietary controllers use an error-squared algorithm.

***Fuzzy Logic Control***    The controllers described heretofore have precise algorithms that produce a precise output to a given set of input conditions. The so-called "fuzzy logic" controller produces an output that is dependent upon the determination of the relative value of the process variable, such as warm, hot, and very hot. In some applications, these controllers can respond to process upsets and process changes more quickly with less overshoot than can a PID controller. Fuzzy logic algorithms and their respective adjustments are proprietary.

## Cascade Control

A cascade control loop consists of two feedback control loops, where each control loop has its own process variable measurement, but only one control loop manipulates the final control element. Although significantly better control can be obtained using cascade control, two process measurements and two controllers are required.

The regulatory controller manipulates the final control element directly (typically varying flow), but the output of the master controller of a cascade control loop (which uses the corresponding regulatory controller process variable as its process variable) sets the set point of the slave controller (which uses the process variable that the final control element varies as its process variable). The slave control loop, typically a relatively fast flow or pressure control loop, will manipulate the final control element to whatever position is necessary to obtain the desired set point. As such, the nonlinearities and hysteresis associated with the valve actuator, process conditions, piping hydraulics, and the like are linearized because the master controller output sets the slave set point, not the output to final control element, as in the case of regulatory control.

Despite the existence of a relatively slow process variable and the relatively fast manipulation of flow by the final control element, regulatory control dictates that the manipulation of both process variables be accomplished with one controller and, hence, one set of tuning constants. Cascade control overcomes this limitation through the use of two control loops, thereby allowing the use of two sets of tuning constants. Using cascade control, the relatively fast loop can be independently tuned for improved control of the fast process variable. Similarly, after the fast control loop is tuned, the slow control loop tuning can be adjusted for improved control because its tuning adjustments do not have to deal with the dynamics of the fast control loop. In addition, each incremental output of the master controller will produce a nearly identical change in the slave controller process variable, making the master control loop easier to tune than its regulatory counterpart, which must deal with valve and process dynamics of the slave process variable. In other words, using individual controllers for the fast and slow control loops separates the tuning of the fast and slow loops into two easier tasks, noting that each controller in the cascade control loop is a regulatory controller that is more directly manipulating the process variable it is attempting to control.

Using a cascade control structure is especially beneficial where changes in the slave controller process variable can upset the process. Because the slave controller quickly returns its process variable to its set point despite nonlinearities and disturbances, master controller output changes does not produce much overshoot, if any. As such, the master controller can better control its process variable by producing smaller fluctuations in the slave process variable sooner than is possible with regulatory control.

It is important that the slave control loop be faster than the master control loop. When the speed of the master and slave controllers are almost equal, the slave control loop may not have sufficient time to reach its set point before the master controller output changes the set point. If this is the case, the control loops will interact and "fight" each other. Detuning the master controller for a slower response may alleviate this problem.

The cascade distillation column bottoms level control shown in Figure 12-4 consists of a flow (slave) control loop, the remote set point (RSP) of which is set by the output of the level (master) control loop. The flow loop is a fast loop that can be tuned for set point changes with minimal overshoot. It is reasonable to expect that the flow would reach and maintain its set point within a few seconds after a set point change. In addition, even though the flow control loop is tuned for set point changes, the flow control loop will react to process disturbances affecting flow much more rapidly than would the regulatory level control loop, which would have to wait until the flow disturbance was seen by the level transmitter.

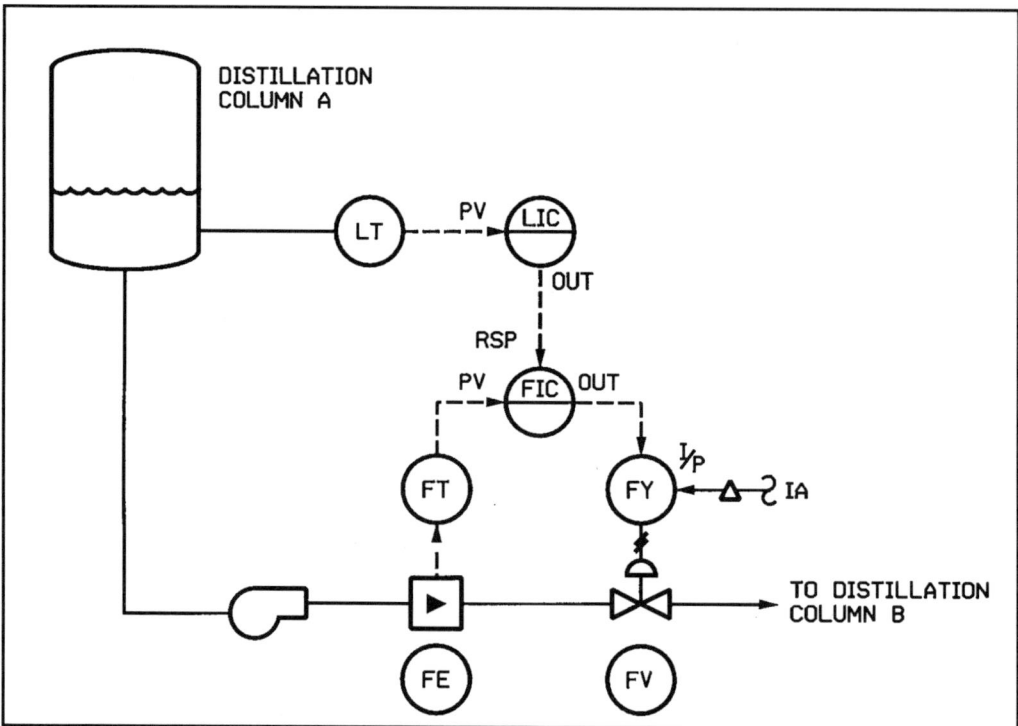

**FIG. 12-4.**    Cascade level control loop (distillation column bottoms)

The level control loop is significantly slower than the flow control loop and should be tuned (after the flow control loop is tuned) for process disturbances, because its set point is not changed often by the operator. The cascade control structure reduces flow variations because the level controller adjusts flow instead of valve position, as was the case of the regulatory control loop. In this manner, flow need only be varied by a small amount to bring the level back to its set point, as compared to its regulatory counterpart, which moved a valve and incremented flow by an unknown amount while measuring its effect on the level.

The advantage of using cascade control is the ability to better control the master process variable by separately and more easily tuning the fast and slow control loops while, at the same time reducing fluctuations of the fast process variable, which in turn can reduce process upsets elsewhere in the process. It should be noted that cascade control can minimize or eliminate the need for the installation of a surge tank between distillation columns—a practice that is not all that uncommon when regulatory control is applied.

# Feedforward Control

Feedforward control is often used in conjunction with feedback control to adjust a controller set point or controller output in response to the change in a related process measurement *before* the effect of the process change is measured by the controller process variable measurement. The intent of feedforward control is to achieve better control of the process variable by having the control system respond to measured disturbances *prior to* the effect of the disturbance being measured at the controller's process variable input, thereby reducing the impact of the disturbance on the process variable and, hence, the process.

Pure feedforward control produces an open-loop response to a disturbance and requires prior quantitative process knowledge, such as the mathematical relationship between process inputs and process outputs. In some applications, pure feedforward control may be difficult to adjust for varying loads and dynamics, so feedforward control is used in conjunction with feedback trim control to trim the process.

***Additive Feedforward Control*** In the application shown in Figure 12-5, additive feedforward control was applied be-

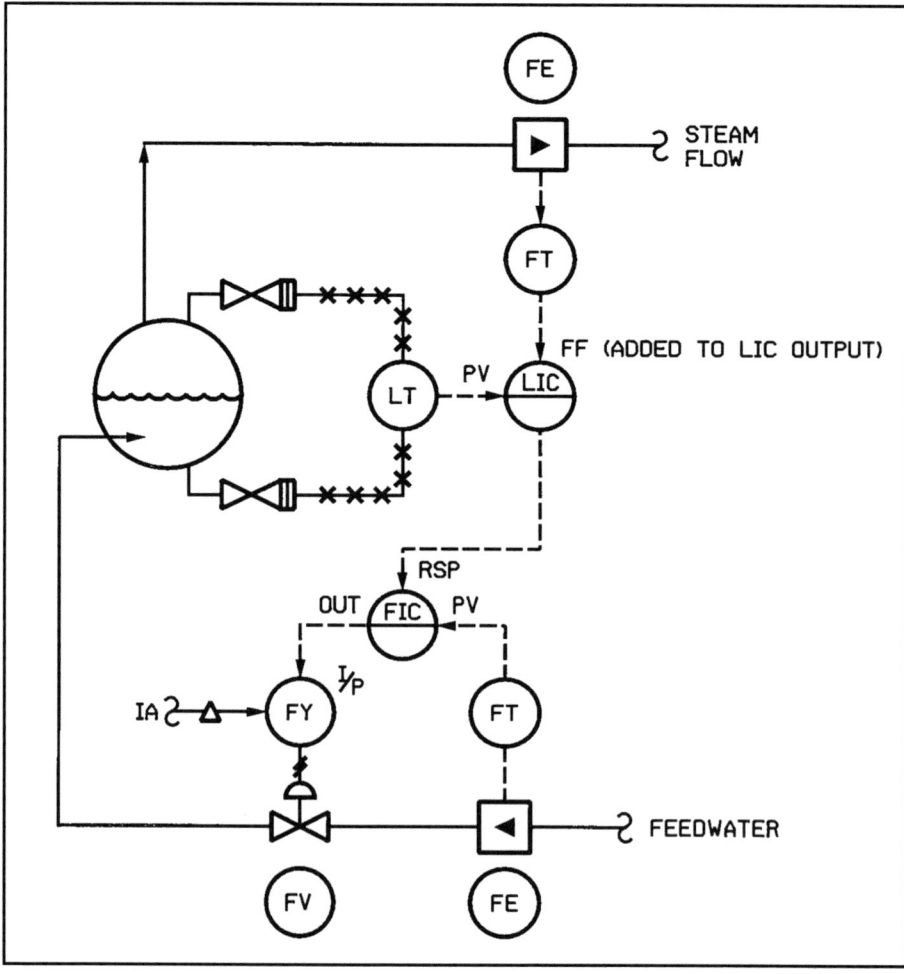

**FIG. 12-5.**  Boiler drum level control loop with feedforward, additive feedback, and cascade control

cause, in the steady state, an increment of the feedforward variable (steam flow) results in a corresponding increment of the manipulated variable (feedwater flow).

The boiler drum level control loop shown consists of a level control loop cascaded to a flow control loop. Due to the fundamental feedback nature of cascade control (that is, the instrumentation will not react until the level changes), abrupt variations in steam demand may cause the drum level to vary significantly. Recognizing that the magnitude of steam flow variations can be related to subsequent level changes, the steam flow measurement can be multiplied by a constant and added to the level controller output (flow controller remote set point) in an attempt to correct for the impending level change.

Note that if cascade control had not been applied, the feedforward signal would adjust the valve directly instead of adjusting the flow set point. If this were the case, a given multiplier may work well at some steaming rates but not at other steaming rates.

***Multiplicative Feedforward Control***  Multiplicative feedforward control can be applied when, in the steady state, an increment of the feedforward variable does not increment the manipulated variable but rather applies to the entire manipulated variable stream. In a well-defined process, feedforward control could be applied in the absence of all other controls to achieve perfect manipulation of the process based upon measurement of process inputs.

An example of multiplicative feedforward control is the ratio control of two flows, only one of which can be manipulated. In the application shown in Figure 12-6, the remote set point of the manipulated flow controller is set in a ratio to the mea-

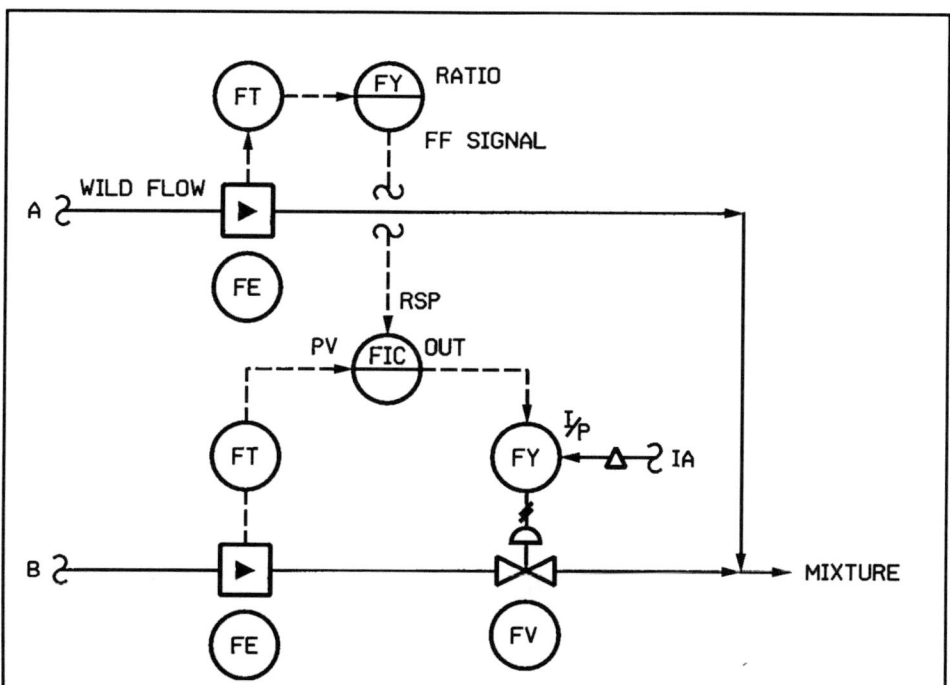

**FIG. 12-6.**  Multiplicative feedforward ratio control loop

sured wild flow. This control strategy would provide reasonable control of the mixed stream.

### Multiplicative Feedforward Control with Feedback Trim
While the use of pure feedforward control is practiced, as in simple combustion controls, the control system cannot compensate for unknown process disturbances, such as composition changes, temperature fluctuations, and the like. Therefore, feedforward control is commonly applied in conjunction with multiplicative feedback trim control to provide the relatively quick feedforward response to measured disturbances and a slower feedback response to unmeasured disturbances.

The simplified boiler combustion control system shown in Figure 12-7, which does not show the cross-ties necessary to lead firing increases (decreases) with air (fuel), illustrates individual combustion air and fuel flow measurements and controllers. The boiler firing rate is changed by adjusting the fuel controller set point. Disregarding the oxygen trim signal, combustion air flow is controlled as a pure feedforward (multiplicative) ratio of the fuel flow measurement.

The purpose of this process is to combust the raw materials (air and fuel) to release heat that is recovered to produce steam. The process produces two products: useful steam and a stack gas that is a waste product. For increased efficiency, it is desirable to recover as much heat as practical from the stack gas. One way to reduce heat losses out the stack is to reduce the amount of excess combustion air that is fed to the burner— a strategy that is not addressed in the pure feedforward control strategy above.

By installing a oxygen analyzer in the stack, the stack gas oxygen content can be measured and fed to the stack oxygen controller, the output of which is multiplied by the fuel measurement to generate the combustion air flow set point. The output of the oxygen controller is used to trim the combustion air flow within prescribed amounts, typically about 10 percent. In this manner, small unmeasured process disturbances (such as the heating value of the fuel, the oxygen content of the combustion air, and the like) can be compensated for using feedback trim. Note that feedforward control is required, despite the installation of the oxygen analyzer, because of the analyzer's sluggish response and potentially poor reliability.

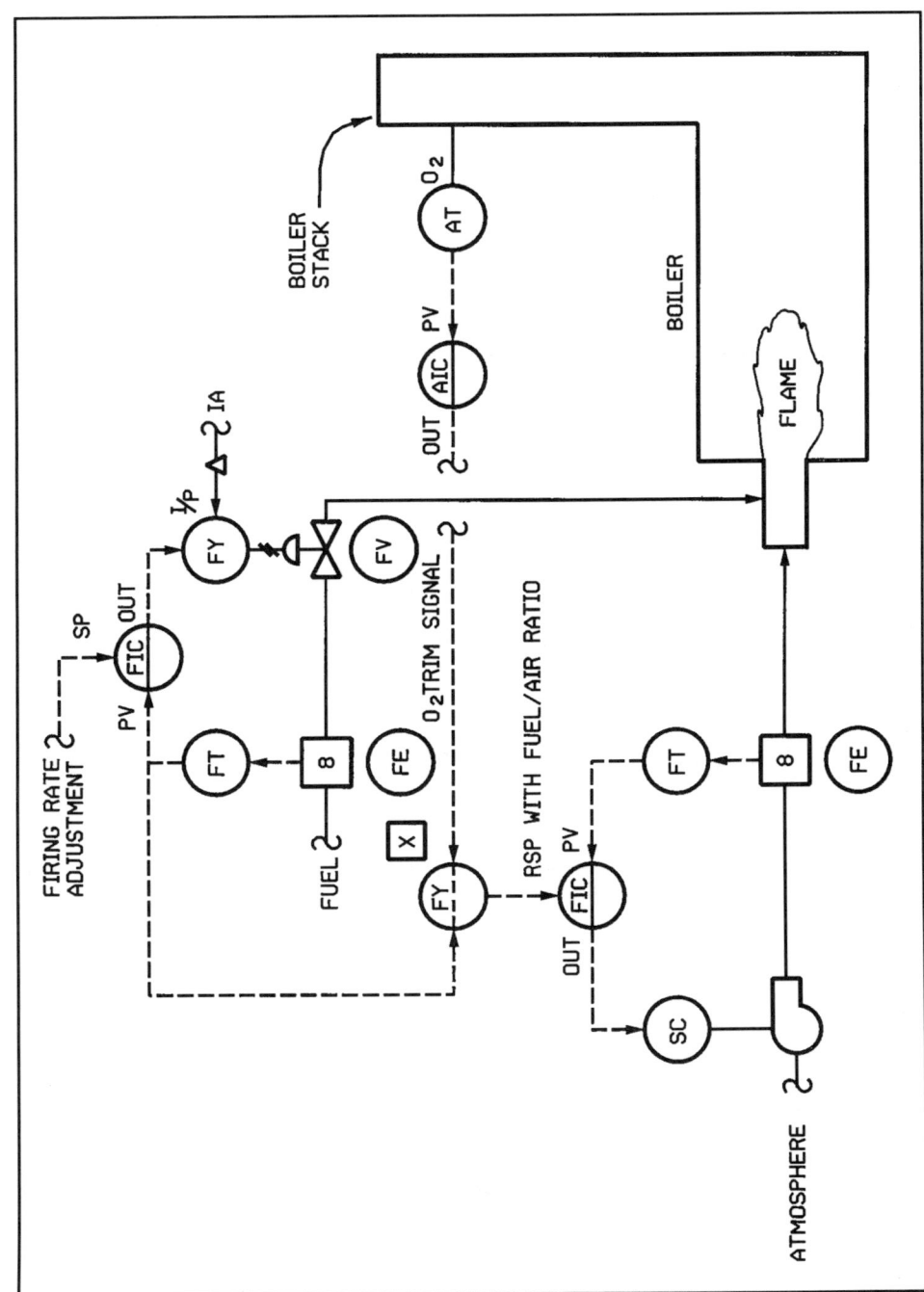

**FIG. 12-7.**   Simplified boiler combustion controls with multiplicative feedforward control and feedback trim

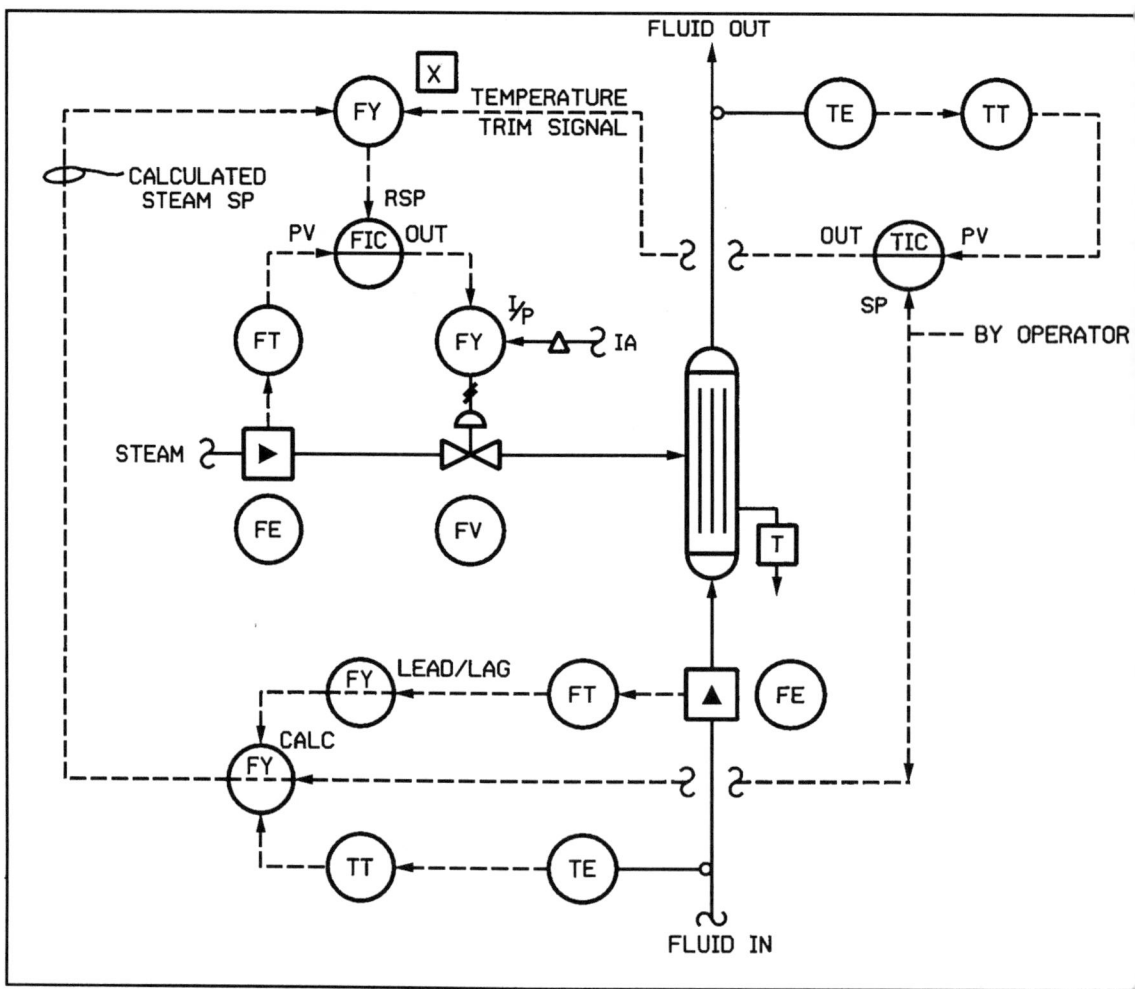

**FIG. 12-8.** Heat exchanger controls with multiplicative feedforward control, feedback trim, and dynamic compensation

***Feedforward Control with Dynamic Compensation*** The feedforward boiler controls described above react relatively quickly to the measured disturbance. In some processes, such as a heat exchanger temperature control, quick response to process measurements can cause process upsets due to the sluggish process. See Figure 12-8.

Disregarding the lead/lag and assuming steady state operation, the fluid flow measurement, fluid inlet temperature measurement, the fluid outlet temperature set point can be used to calculate the steady-state amount of steam required to raise the fluid exit temperature to its set point. In the steady state, the calculated steam flow can be used as the steam flow controller set point. The temperature trim controller is used to trim the calculated steam flow set point (using feedback) to take heat losses, steam header pressure variations, and the like into account.

It should be noted that, although measurements occur relatively quickly, the process can exhibit a delayed dynamic response to flow changes; therefore, it is not desirable to immediately change the steam flow controller set point based upon current measurement values. Because changes to fluid flow would abruptly (multiplicatively) affect steam flow, the lead/lag is used to modify the fluid flow signal being fed to the calculation to temper the response of the control system to fluid flow changes.

# Feedforward Action

Feedforward action is the utilization of process interaction to affect feedforward control *without* feedforward measurements. This differs from feedforward control where measurements were used to allow the control system to respond to process disturbances *before* the effects of the measured disturbances affected the controlled variable. Applying feedforward action requires an in-depth (but not necessarily mathematical) understanding of the process, process equipment, and their dynamics.

The relatively simple heat exchanger control application of Figure 12-9 implements a deceptively complex control system. The temperature measurement and controller are straightforward, but steam is manipulated using a remotely set pressure regulator. The remotely set pressure regulator implements a cascade control strategy by varying steam pressure in proportion to the temperature controller output (instead of the usual handwheel or set screw). In addition, should the inlet fluid temperature decrease (increase) or the fluid flow increase (decrease), more steam will condense, causing the pressure to fall and the regulator to open in order to compensate for the load almost immediately after it is sensed by the steam *in the heat exchanger* and minimizing the effect of the process change on the outlet temperature. In addition, the pressure regulator minimizes the effects of steam header pressure variations on the fluid outlet temperature. The steam pressure measurement is installed for diagnostic purposes.

In another application, the purpose of the makeup air flow control shown in Figure 12-10, is to provide a constant air

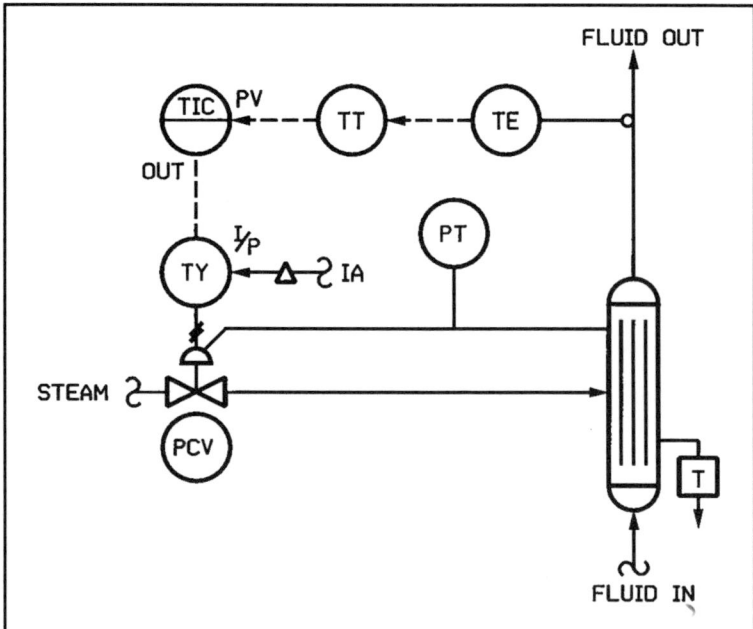

**FIG. 12-9.**     Heat exchanger controls with feedforward action

flow to the downstream load. This control strategy appears to be similar to the feedforward ratio control described above, except that the controlled air set point is the difference between the desired air flow and the "wild" air flow, instead of a ratio. One subtle difference between the two loops is the use of a variable speed drive to vary fan speed to manipulate air flow, an equipment selection decision that was no accident.

Were the controlled air manipulated by controlling a valve or damper, the wild air flow changes would have only a small effect on the controlled air flow during transients and significantly affect the outlet air flow. Stable air flow would seem to be a virtue, however, variations in the wild flow would cause air flow controller set point changes that (for hydraulic reasons) would result in the movement of the control valve or damper. These variations would also cause the outlet air flow to vary until sufficient time had elapsed for the air flow control loop setpoint to stabilize and for the air flow to reach its set point.

Maintaining a forward flow with a variable speed drive controlled system (hydraulically) positions the relatively flat portion of the fan curve such that wild flow variations cause the fan, in response, to ride out on the curve, increasing air flow

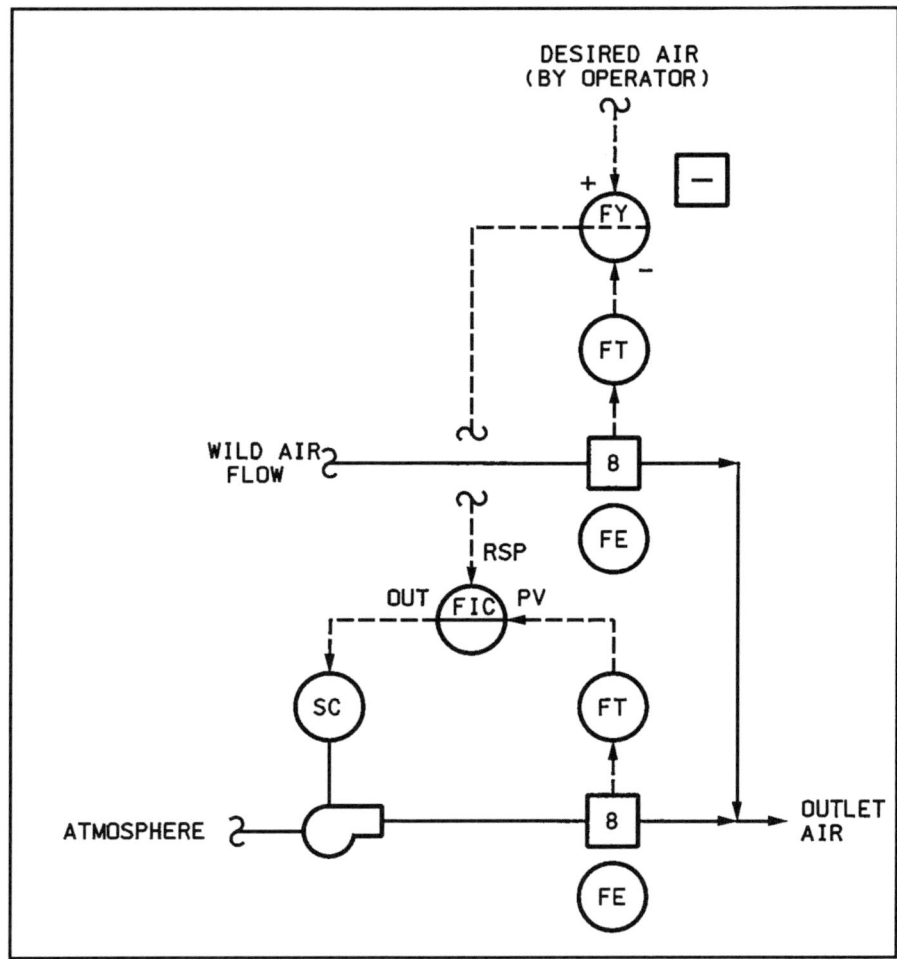

**FIG. 12-10.** Makeup air flow control with feedforward action

with no appreciable change in fan speed. Because of this feature, this system will control the total flow better in the face of more abrupt wild flow variations because the feedforward action of moving on the fan curve occurs through the process. In addition, fan operation is more energy efficient.

## Constraint and Override Control

The control loop may be allowed to manipulate the final control element freely, as long as a particular constraint is within limits. However, when the constraint remains outside these limits, a controller may become active and assume control of the final control element by either directly manipulating the final control element or overriding the control loop, perhaps using a signal selector.

## Summary

Advanced regulatory control provide a powerful set of tools to improve control of the process. Seemingly straight-forward, these tools can be combined to produce results that are strikingly superior to the results achievable using regulatory control.

# Chapter 13

# Advanced Regulatory Control Design and Implementation

Advanced regulatory control should be designed from the top down and implemented from the bottom up. Failure to use this implementation strategy may reduce the effectiveness of the installed control system.

## Design

Top-down design (see Figure 13-1) implies that the detailed specifications of a particular instrument (bottom level, entailing details) cannot be implemented until the process control goals and control strategies (top level, containing more general ideas) are developed and finalized. Similarly, a number of intermediate steps, which may overlap somewhat between the top and bottom, should be performed, starting from the most general and constantly moving to the more specific.

***Define the Product and Its Quality*** Examination of a process flow sheet or process and instrumentation diagrams (P&IDs) will usually result in a list of flows, levels, pressures, and temperatures to be controlled in order to manage the process. Implementation based upon this list will result in an installation

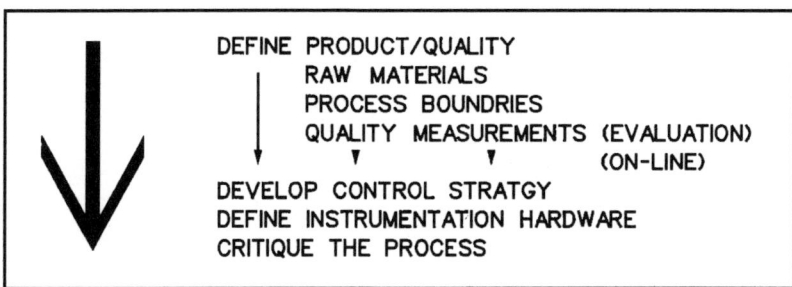

**FIG. 13-1.** Top-down design

that neither streamlines the process, making it more efficient, nor exploits available control technology to improve control of the process.

Before attempting to instrument the process, the output of the process should be identified and its quality defined. *THE PURPOSE OF THE PROCESS SHOULD BE CLEARLY UNDERSTOOD.* Quality may include a number of factors that include peak flow, average flow, minimum flow, flow stability, composition, color, texture, temperature, pressure, and the like.

For example, a 99 percent pure chemical produced at a nearly constant flow rate with less than 0.5 percent of a given impurity could be the specification of a chemical intermediate produced by the process to be controlled, even though the intermediate is used in the next process and not sold. On the other hand, maintaining the level of a tank perfectly stable is not a product, despite its potentially critical nature in the process.

***Define the Raw Materials*** Having defined the product of the process, the raw materials used to make the product should be defined and their quality quantified. Raw materials to a given process can include chemicals, materials, utilities, process equipment, labor, and other items needed to produce the product.

For example, utilities such as steam are often taken for granted. However, if the steam generation and distribution systems are not "stiff," soot blows, rain on uninsulated pipes, and the like can reduce steam pressure and upset the process. Such knowledge could trigger the consideration of pressure compensation for steam flow control measurements.

***Define the Boundaries of the Process*** Because control problems can stem from any of these raw materials, the boundaries of the project should include systems that affect all of the raw materials, *even if improvements to these systems will not be currently funded by management.* This approach is preferred to avoid unknowingly causing the control system to hide a real problem that will be more difficult and costly to correct later. Often, subtle changes in the operation of a process can have dramatic positive or negative effects on upstream, downstream, and related processes.

For example, investigation may show that the boiler steam pressure controller may be improperly adjusted or defective.

In the absence of this information, installing pressure compensation may alleviate the problem at hand, but the underlying problem may adversely affect other processes. Treating the problem does not preclude pressure compensation because even with a "constant pressure" steam supply, other considerations, such as piping losses and the effects of multiple users, can affect the process when temperature compensation is not applied. In general, improved process performance can be achieved when variations in raw material uniformity and quality are reduced.

***Define Quality Measurements for Evaluation***  Recognizing that product quality and capacity are important, which measurements to make to determine quality and where to make those measurements should be defined for evaluation. Likewise, measurements of in-process material should be defined. By performing this step, control of the process to achieve final product quality can be broken up into smaller processes that are usually easier to address.

For example, producing 99 percent pure material may consist of a reaction that produces 50 percent pure material, followed by a concentration process that produces 99 percent pure material. This problem may now be broken into two simpler problems: appropriate assays may show that the reacted material varies between 30 and 60 percent purity, and the product varies between 98 and 99.5 percent purity. These components can now be addressed individually, recognizing that reducing the variations in composition from the reactor (using an in-process quality measurement) may obviate the need for upgrading the concentration process.

***Define Quality Measurements for On-Line Operation***  When the evaluation of the data is complete, on-line process quality measurements can be defined at various locations in the process, if possible and practical. These measurements are important, because the purpose of the control system will be to stabilize these measurements. In addition, these measurements can be viewed in real time, which allows timely correction that can significantly improve the operation of the process.

***Develop the Control Strategy***  The control strategy should be developed to stabilize and control the on-line quality measurements, utilizing the advanced regulatory control tools available.

*The strategy should measure and control as close as possible to the process to minimize dead time* . For example, if vapor flow is important and reacts quickly to process changes, the flow should be measured instead of waiting tens of minutes for a downstream condensation tank level to rise. Measurements such as these make slow loops into faster loops that can react more quickly to process upsets, thus making the process easier to control.

Define the process variables that should be managed, being sure to include calculated measurements based upon more than one process variable. Despite the number of flow measurements and valves, flow measurements are not process variables that are usually managed; flows are usually manipulated or used in calculations. The managed process variables should be paired with an appropriate manipulated variable—typically a flow.

Understand that cascading slow loops into fast loops will tend to stabilize both the controlled variable and the manipulated variable. *Improved process stabilization may require that virtually all flows be controlled as the slave in a cascade control loop.*

Fast loops should manipulate and control one process variable only. For example, if it is desired to control the sum of two flows where only one can be manipulated, the flow controller should manipulate the flow directly, and the flow controller set point should be calculated by subtracting the wild flow from the desired flow. In other words, do not build additional process dynamics into process variables that will be detrimental to loop tuning.

**Define Instrumentation Hardware**  The function and location of field instruments performing the measurements needed to implement the control strategy should be defined. The control instrumentation should also be defined. This information can be used to estimate the project.

**Critique the Process**  When the control system has been developed, the control system and process should be examined to determine where improvements can be made. For example, the process engineer may include equipment that requires additional instrumentation to manage, but which may not seem necessary given the availability of recently introduced instrumentation. Location of a piece of equipment on the next higher platform may provide sufficient head for gravity flow, eliminating the need for a pump while allowing more space on the lower platform for maintenance. The implementation of some measurements may require that the process be modified.

All proposed improvements should be evaluated for their technical merit to avoid missing a technical opportunity for improvement or causing costly changes during implementation.

## Implementation

Bottom-up implementation (see Figure 13-2) implies that the details regarding the overall process strategy (top-level control) cannot be implemented until the details regarding the process equipment and field instrumentation (bottom-level measurement and manipulation) are developed and finalized. Similarly, a number of intermediate steps that may overlap somewhat between the bottom and the top should be performed, starting from the specific and constantly moving to the more general. Implementation should be based upon the information developed during the design phase of the project.

***Provide Equipment Data***  Information such as openings on tanks, equipment nozzle data, and the like, are required by the equipment manufacturer in order to fabricate the equipment. Due to the long lead times involved as compared to instrumentation, this information must be provided to the appropriate party relatively early in the project.

***Specify Hardware***  The field equipment and control hardware should be specified to provide timely delivery and to allow sufficient time for software development. Field equipment should be specified to measure the *desired* process measurement.

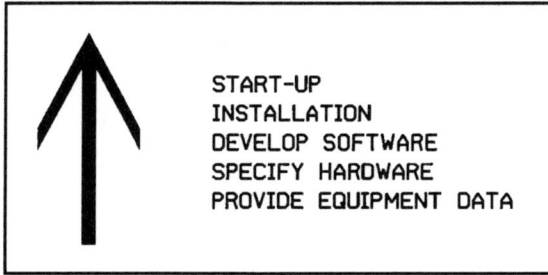

**FIG. 13-2.**   Bottom-up implementation

***Develop Software***  Software development to implement the control strategy should proceed from the bottom up, that is, starting with the definition of inputs and outputs and progressing to fast control loops, slow control loops, and beyond. *Care must be taken to make sure that the operator interface is easy to use and will achieve bumpless, balanceless transfer between the various controller operating modes, such as manual, automatic, cascade, hold, tracking, and the like.*

***Installation***  Careful attention must be paid to the installation details to ensure that the instrument is installed correctly and measures the desired process variable at the proper location without introducing delay.

***Start-up***  should proceed from the bottom up by verifying the process equipment, field instrumentation, fast control loops, slow control loops, and beyond. Fast loops are usually tuned for set point changes; slow loops are usually tuned for process disturbances.

## Summary

The implementation of a successful advanced regulatory control system entails top-down design and bottom-up implementation, that is, design should start from the general and migrate to the details, whereas implementation should start with detail and migrate to the general. Violation of these guidelines can destine projects to failure before project activities commence.

## ❖ Chapter 14

# *Applying Advanced Regulatory Control*

To illustrate the application of advanced regulatory control, the stripping process in Figure. 14-1 is considered. The intent of this exercise is to illustrate the implementation of the design process proposed in the previous chapter. Due to the numerous process constraints and idiosyncracies, the

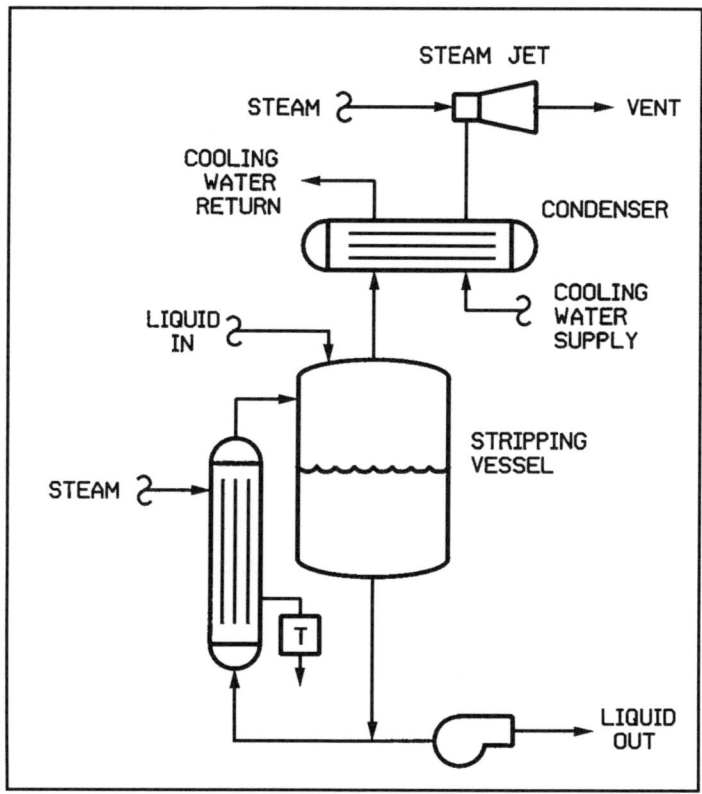

**FIG.. 14-1.** Stripping process

proposed control system should not be construed as a viable or an appropriate way to control a particular process.

The process consists of a vessel under vacuum (provided by a steam jet) in which the liquid is boiled, using a thermosyphon reboiler. The vapor is condensed in a water-cooled condenser that allows volatile material to exit the system via the vent. A pump is used to transfer the liquid to another part of the process.

The top-down design procedure can be used to develop a control strategy.

## Define the Product and Its Quality

The purpose of this process is to produce a liquid that contains only small amounts of volatile material. The quantity of the material processed must approximately equal the flow of liquid from the upstream process, which contains small amounts of volatiles. In addition, the process produces a waste stream that goes to vent, warm cooling water, and steam condensate. It is desired that the amount of nonvolatile material in the vent steam be minimized.

## Define the Raw Materials

The raw materials of this process are the liquid with unknown amounts of volatile material, cooling water, steam to the reboiler, and steam to the jet. Steam pressure can vary by about 10 percent of the absolute steam pressure. Cooling water temperature will vary with time of day and season; cooling water flow can vary because of occasional manual balancing by the operator.

## Define the Boundaries of the Process

The process is ostensibly limited to the process shown on the sketch; however, the upstream liquid process, downstream liquid process, cooling water supply, cooling water return, steam supplies, and condensate return systems will not be considered off limits.

## Define Quality Measurements for Evaluation

The quality of the product is determined by the quantity of volatile materials in the liquid leaving the process. Sampling this liquid and analyzing for volatile material would be a measure of product quality.

# Define Quality Measurements for On-Line Operation

Installing an on-line analyzer system would be prohibitively expensive and is deemed impractical in this application. However, it is known that if the liquid in the vessel is maintained above a certain temperature at a given vacuum, the liquid exiting the process will contain a sufficiently low amount of volatile materials.

It would seem that the temperature of the liquid in the vessel can be used as an indication of the on-line quality of the liquid, an approach that would seem reasonable if the vacuum were maintained constant at the given vacuum. Because the vacuum can vary significantly, the temperature measurement (which is really intended to infer composition) should be (mathematically) corrected for changing vacuum, using an absolute pressure measurement. In this manner, composition can be inferred more accurately than by using temperature alone to infer composition (See Figure 14-2).

# Develop the Control Strategy

The control strategy illustrated in Figure. 14-2 can be developed starting with the loops "closest" to the process. Because virtually all controllers that operate final control elements manipulate flow in order to control another parameter, the application of flow controls should be considered first, followed by the management of volumes, and finally, the control of composition.

As was previously stated, most manipulated variables are flows, which in this application include:

- liquid flow to the next process,
- steam flow to the jet, and
- steam flow to the reboiler.

The cooling water flow to the condenser is an additional, manipulated variable that is ignored because the cooling water will be manually balanced.

It is assumed that the liquid flow into the process cannot be manipulated, but the control of the upstream process should be examined to determine whether improvements can be made to steady this flow.

Variables that must be controlled and/or managed include:

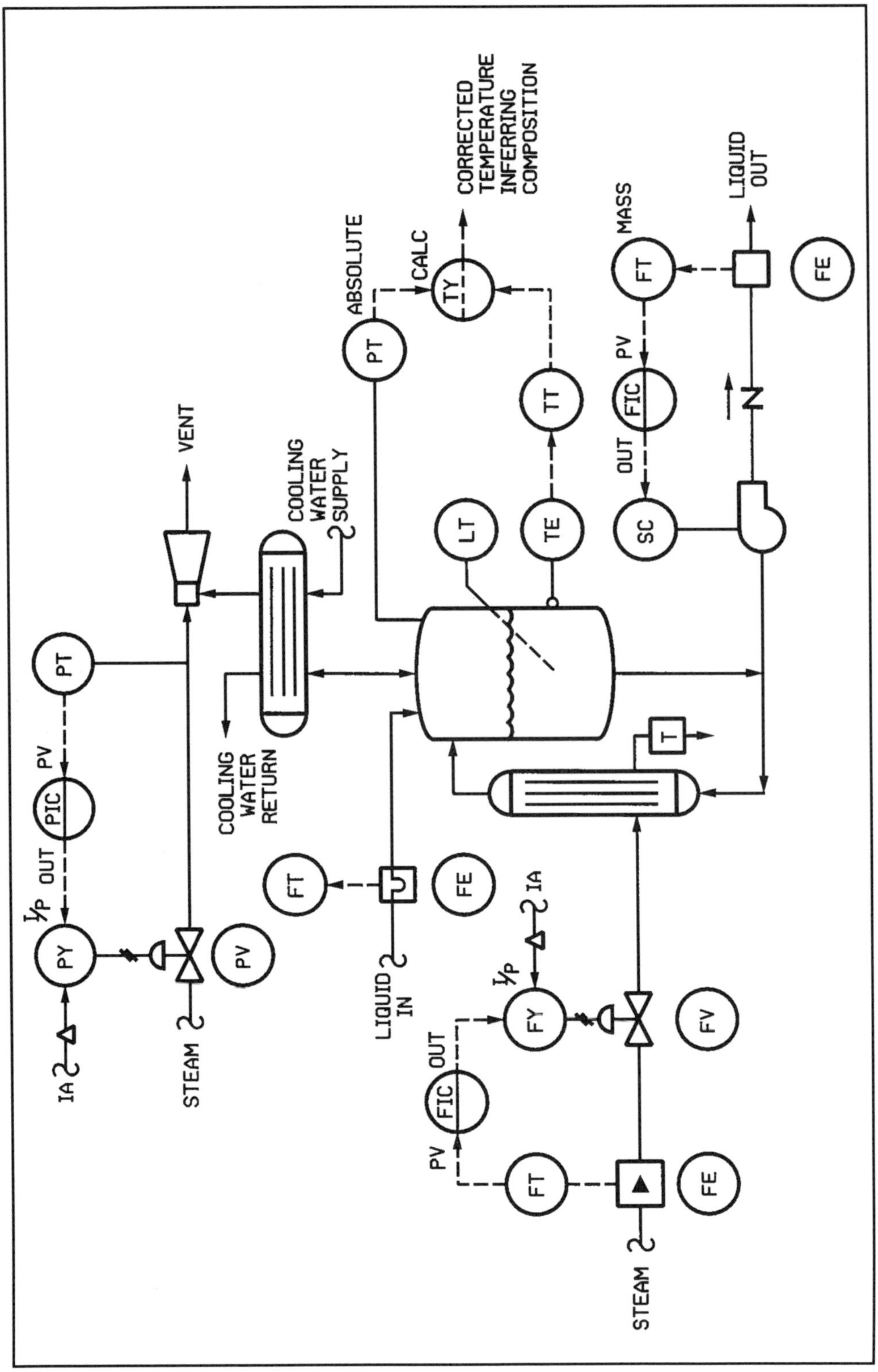

**FIG.. 14-2.** Stripping process with fast control loops and controlled variable measurements

- vessel liquid level,

- vessel vacuum, and

- liquid composition in the vessel.

The condenser outlet temperature is ignored, because it is assumed that the operator will balance the system with the proper amount of cooling water.

Ideally, manipulating one variable will affect only one controlled variable; however, analysis of the process will reveal that this is not the case in this application. For example, consider that a steam flow increase to the reboiler will increase the temperature of the liquid, causing more boiling, which in turn increases the absolute pressure in the vessel and lowers the liquid level.

To stabilize and linearize the manipulated variables, at first glance it would appear that flow control loops should be applied to all flows that can be manipulated, because these loops are fast process variables that will react significantly faster than absolute pressure or composition. However, steam jet suction is determined by steam pressure, which reacts quickly to valve manipulation and can be considered a fast control loop. In this application, steam pressure will be controlled instead of steam flow.

Vessel temperature should be measured to infer composition. The incoming liquid flow should be measured even though it cannot be controlled.

***Vessel Liquid Level*** As alluded to above, the liquid level in the vessel can be affected by changing any of the manipulated variables, but it is most directly affected by the liquid flow leaving the vessel. This level should be controlled at a level slightly above the top of the reboiler tube sheet in order to utilize the full heat exchanger surface and maintain steady recirculation.

To control the level and stabilize the flow to the next process, a level control loop cascading level into flow would seem appropriate (see Figure. 14-3). It is also appropriate to investigate the control loop (if any) for the liquid that feeds this process, because its stabilization (perhaps using cascade control) will reduce the fluctuation of the liquid flow leaving the vessel.

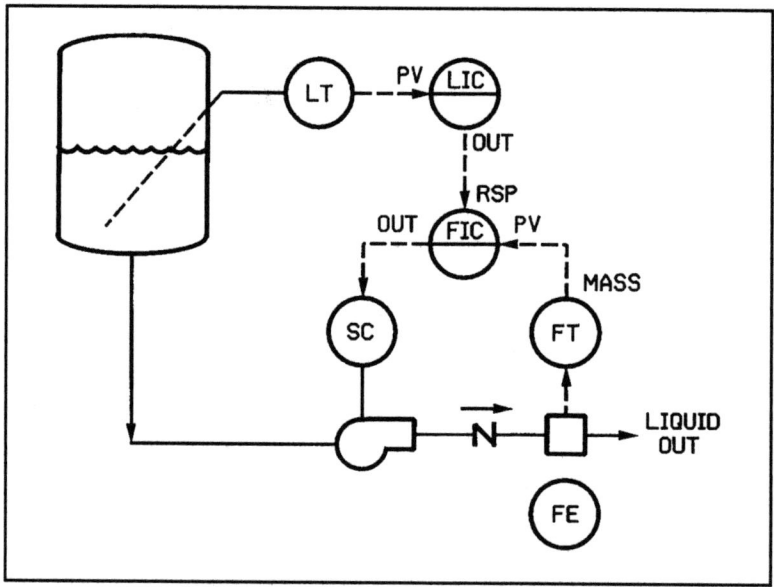

**FIG.. 14-3.**   Vessel liquid level control

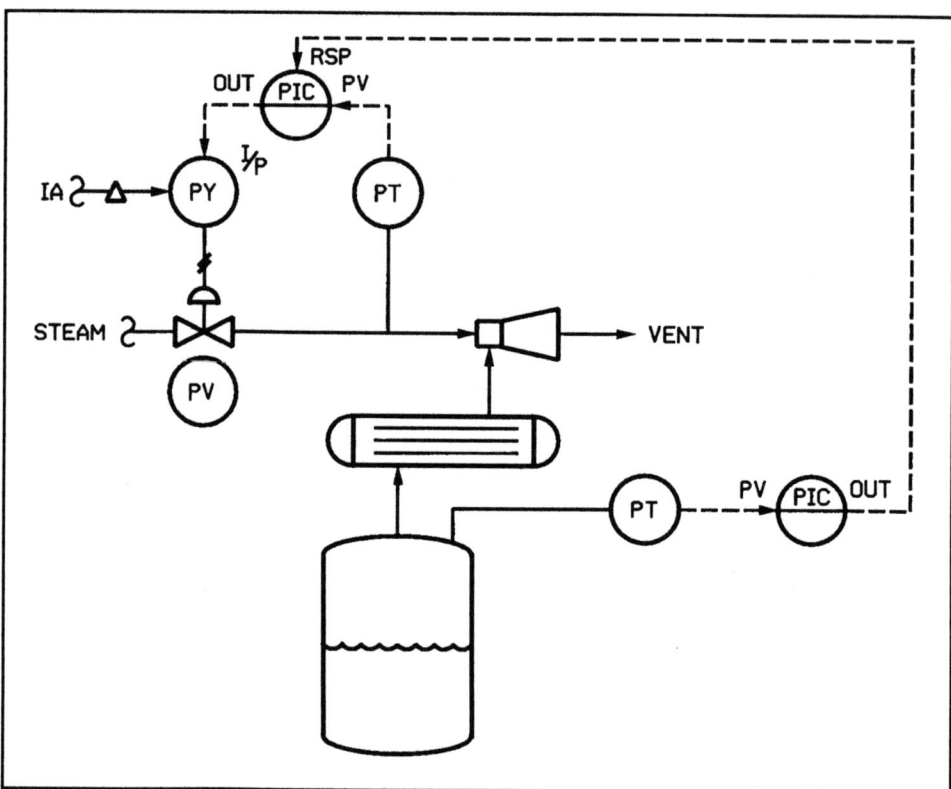

**FIG.. 14-4.**   Vessel vacuum control

***Vessel Vacuum***  Vessel vacuum is indirectly affected by reboiler steam flow, somewhat more affected by liquid flow into the vessel (which cannot be manipulated), but much more directly manipulated by adjusting the vacuum source. An absolute pressure controller cascaded to the steam jet pressure controller can be applied to control vacuum in the vessel. See Figure. 14-4.

The steam pressure control loop can be more economically and more simply implemented using a remotely set pressure regulator.

***Liquid Composition in the Vessel***  Based upon process knowledge, the measured liquid temperature in the vessel can be corrected to a reference absolute pressure by calculating the temperature at the reference absolute pressure condition that would have the same percent volatile material as the actual liquid at the measured temperature and absolute pressure. In this manner, controlling the corrected temperature will control the liquid composition in the vessel and compensate for vessel pressure fluctuations.

Vessel temperature is most affected by steam to the reboiler, therefore, a composition controller cascaded to reboiler steam flow can be applied (see Figure. 14-5). Pressure compensation that could be applied to the steam flow measurement to compensate for steam header pressure disturbances is not shown.

***Feedforward Control***  Feedforward control may be needed to allow the control loops to react to load changes, that is, changes to the liquid flow entering the vessel. If not corrected for in the steady state, these changes will reduce the vacuum, lower the liquid temperature, and increase the liquid level. Therefore, in the steady state, both steam flows and the liquid flow leaving the vessel must be increased.

The cascade absolute pressure control loop will probably react relatively quickly to a load change through its feedback algorithm, because the load change will almost immediately affect the vessel absolute pressure measurement. The outputs of the level controller and the corrected temperature controller may require additive feedforward. However, if the reboiler steam control loop were replaced with a control loop that used feedforward action (using a remotely set pressure regulator), the corrected temperature controller would not require feedforward control.

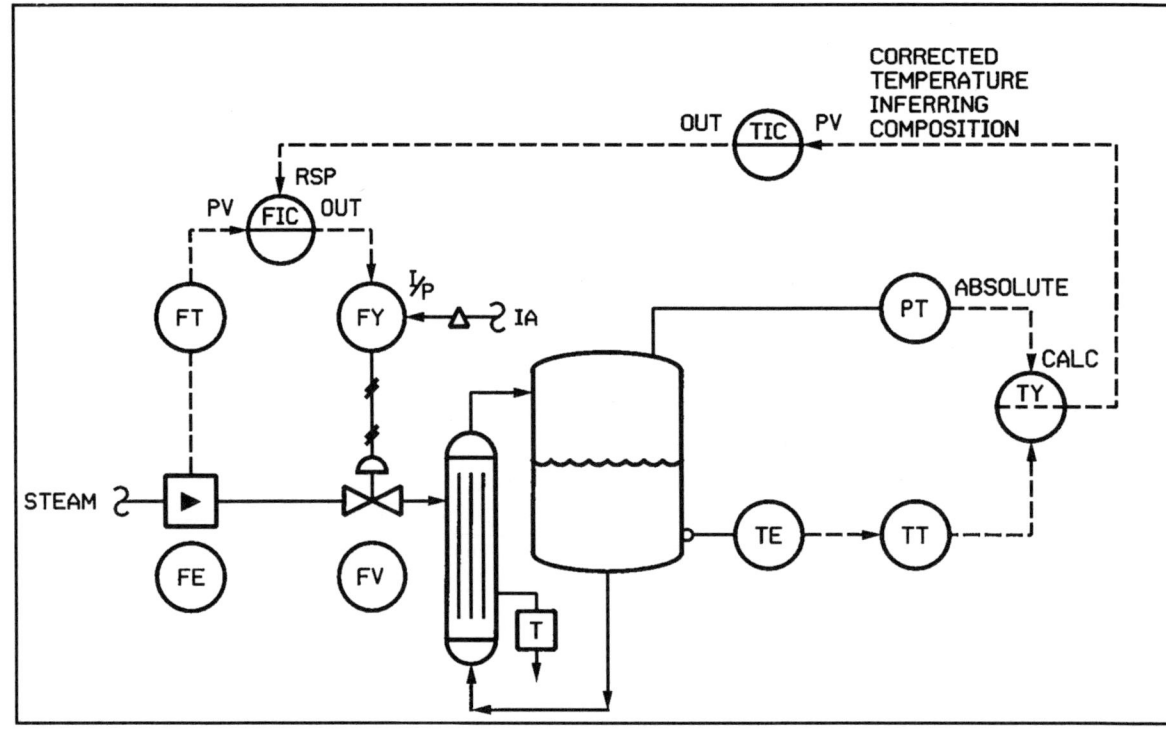

**FIG.. 14-5.**   Liquid temperature control

***Dynamic Compensation*** As mentioned above, the absolute pressure control loop should be relatively responsive to pressure changes and should not require dynamic compensation. However, the response of the corrected temperature controller may benefit from dynamic compensation of the feedforward signal (when feedforward action is not used). Dynamic compensation of the level control loop feedforward signal may be required (see Figure. 14-6).

***Logic and Data Manipulation*** A controller will effectively lose control of the process when its output is at its minimum or maximum value. Override control can be used in some applications to enable other control loops that will attempt to keep the process under control. For example, if the reboiler steam control valve is completely open, a corrected temperature override controller can be activated that will generate the set point of the steam jet pressure controller in an attempt to maintain liquid composition by increasing steam jet pressure.

Constraint control can also be used to meet certain control objectives by relaxing other control objectives. For example, if

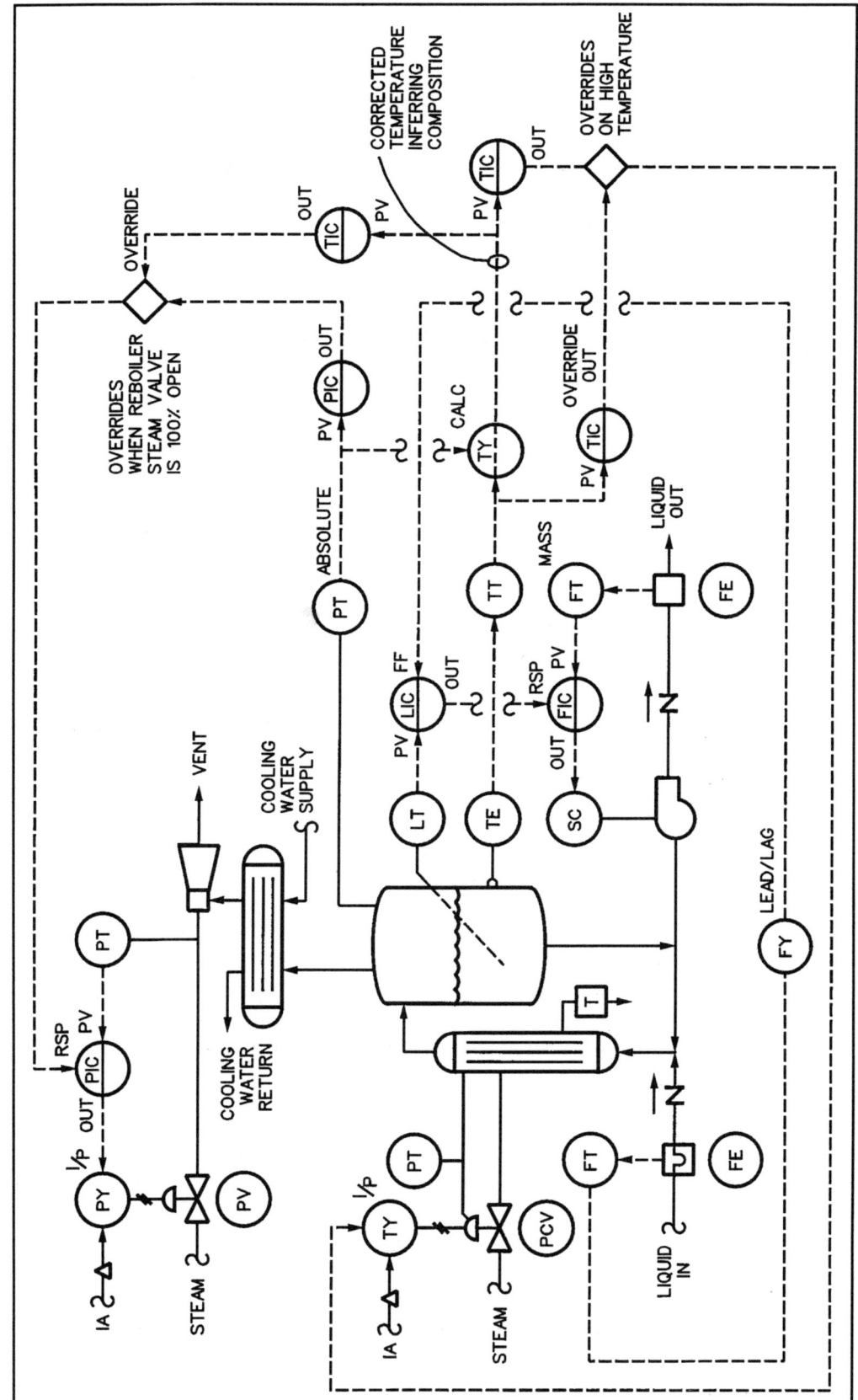

**FIG.. 14-6.** Stripping process with feedforward action, feedforward control, dynamic compensation, override control, constraint control, and liquid fed to the inlet of the reboiler

the liquid could not be heated above a certain temperature without appreciable degradation, a temperature constraint controller may become active and set the reboiler steam flow to maintain the maximum allowable temperature, at the expense of the composition control objective. In an attempt to compensate for the relaxation of the composition control objective, process conditions may also activate the corrected temperature override controller.

## Define Instrumentation Hardware

Knowledge of the process and the control strategy developed above can be used to define the instrumentation necessary to implement the control strategy. In this application, field instrumentation suitable for hazardous electrical locations and an industrial environment is necessary. The controllers or control system must be capable of implementing the above cited algorithms and provide flexibility for future change.

## Critique the Process

As stated above, the upstream process should be examined to determine if the incoming flow can be stabilized or, better yet, manipulated in an attempt to stabilize the flow leaving this process.

A surge tank may be installed between this process and the downstream process. If the liquid flow leaving this process can be adequately stabilized, the surge tank may be eliminated or, better yet, not installed.

Introduction of the liquid at the reboiler inlet may dampen the effect of load changes on vessel pressure and temperature. This would be especially beneficial when feedforward action is used, because the reboiler will modulate the steam valve quickly in response to the load change *before* the liquid reaches the vessel.

## Summary

As can be seen from the above analysis, control system design takes many facets of the operation into account. In addition, even the experts will not agree upon one correct control system design for a process. However, proper application of advanced regulatory control can alleviate the need for more sophisticated techniques in all but the most demanding control applications.

# Index